FROM PIGEONS TO NEWS PORTALS

Media & Public Affairs

ROBERT MANN, SERIES EDITOR

Media & Public Affairs, a book series published by Louisiana State University Press and the Reilly Center for Media & Public Affairs at the Manship School of Mass Communication, LSU, explores the complex relationship between knowledge and power in our democracy. Books in this series examine what citizens and public officials know, where they get their information, and how they use that information to act. For more information, visit www.lsu.edu/mpabookseries.

# FROM PIGEONS TO NEWS PORTALS

## FOREIGN REPORTING AND THE CHALLENGE OF NEW TECHNOLOGY

EDITED BY DAVID D. PERLMUTTER
AND JOHN MAXWELL HAMILTON

Louisiana State University Press
Baton Rouge

Published by Louisiana State University Press

Manufactured in the United States of America
First printing

*Designer:* Michelle A. Neustrom
*Typeface:* Adobe Garamond Pro, Bank Gothic
*Printer and binder:* Thomson-Shore, Inc.

*Library of Congress Cataloging-in-Publication Data*

From pigeons to news portals : foreign reporting and the challenge of new technology / edited by David D. Perlmutter and John Maxwell Hamilton.
p. cm. — (Media & public affairs)
Includes bibliographical references and index.
ISBN 978-0-8071-3280-7 (cloth : alk. paper) — ISBN 978-0-8071-3282-1 (pbk. : alk. paper)
1. Foreign news. 2. Journalism—Technological innovations. I. Perlmutter, David D., 1962– II. Hamilton, John Maxwell.
PN4784.F6F76 2007
070.4'332—dc22

2007017914

The paper in this book meets the guidelines for permanence and durability of the Committee on Production Guidelines for Book Longevity of the Council on Library Resources. ♾

For Dee Dee and Kevin Reilly, in appreciation for their vision and all they have done to advance the study of media and public affairs through the Reilly Center at the Manship School.

# CONTENTS

# FROM PIGEONS TO NEWS PORTALS

# 1

# INTRODUCTION

## *The Challenge of Technological Change in Foreign Affairs Reporting*

DAVID D. PERLMUTTER AND JOHN MAXWELL HAMILTON

We live in an age where, as Thomas M. Disch put it, science fiction is the "dreams our stuff is made of." Every year brings a host of new gadgets that change the methods of communicating with each other and how we interact with our world. Perhaps as remarkable is how quickly innovative technology, from ATMs to podcasting, is integrated into our lives, or at least those of the true avant-garde of the high-tech age, young people and professionals. In the case of the former, a recent study reported that upwards of 80 percent of fifteen-year-olds use instant messaging and a majority have their own Weblog. The latter, journalists, have in just the past decade wielded new devices that significantly alter the way they cover and present news. For example, seventy to seventy-five cases of gear were needed for a television crew to cover the Afghanistan war in 2002.[1] Eighteen months later, when America invaded Iraq, a television correspondent required only seven cases to transmit a report, live, to the United States.

While we may now readily accept that reporters can broadcast from satellite-connected and video-enabled cell phones hanging on their belts, it is quite another thing to grasp the implications. These are elusive. They unfold slowly, with many false starts and considerable resistance even from those who use the technology most. Interestingly, many of the journalists who are quick to deploy new generations of equipment complain they are doing their jobs less well. Policy makers fret that our wired world is more difficult to manage. Audiences may feel overwhelmed by the data explosion.

This book is an exploration of implications. In it, we examine ways that people have used radical and new media technology—from satellites and cell phones to digital convergence and the Internet—to affect the creation and content, economics and logistics, delivery modes and venues, amount and style of coverage, and accuracy and reliability of foreign news and in turn its influences on public opinion and government policy making. The emphasis will be on what is happening now, with an attempt to predict future trends. Our focus is America, for the practical reason of needing to limit the scope of the book to manageable proportions and because we are interested in the issues raised by media technology for the special set of conditions surrounding the United States' place in the world.

Our conceit is to assess the validity of basic "truths" long held about foreign affairs and the news media. Every U.S. journalism school teaches a similar set of skills, codes, values, assumptions, and standards that define and delimit the ethical and successful news worker, whether domestic or foreign correspondent. Likewise, similar assumptions govern theory and practice among professionals—in media or in public affairs work, in government or among nongovernmental organizations (NGOs). We think, however, that both independent amateurs—from bloggers to eyewitnesses with their cell phone cameras—and news workers alike are overturning many of our social, economic, and managerial paradigms. We hope here to survey, appreciate, and demystify the new foreign correspondence.

We are not, however, technological Pollyannas, assuming that new media are always the deliverers of virtue and wisdom. Technological innovation is inevitable, and mastering that change for the better requires asking tough questions. It is not enough simply to register that, for example, the lighter toolkit of modern foreign correspondents makes them veritable one-person, "live from ground zero" news networks. We want to explain how the *minds* of news workers are evolving in line with their gadgetry. Likewise, it is not sufficient to simply document that ordinary citizens can now, via their cell phones and blogs, become "citizen foreign reporters" with worldwide audiences. We show how such "new news" mechanisms affect the way the public *conceives* of an entity called "foreign news."

Thomas Paine once proclaimed that the American Revolution was not just a political upheaval but a metaphysical leap: "Our style and manner of thinking have undergone a revolution more extraordinary, than the political revolution of the country. We see with other eyes; we hear with other ears; and think with other thoughts, than those we formerly used."[2] We think a cognitive revolution is indeed taking place in the way people think about and use news from foreign lands, and even how they define what is "foreign." Accordingly, in each chapter to follow, leading academic and professional experts recast and challenge a commonplace of foreign news factually and normatively and speculate on what the future holds.

## A BRIEF HISTORY OF THE EVOLUTION OF TECHNOLOGY

We begin in this chapter with some perspective. First, the notion of foreign correspondence is a relatively new one. For millions of years human beings and the human mind evolved within small hunter-gatherer groups and tribal societies, where the furthest necessary knowledge of any utility was that mammoths had been sighted in the next valley, that there was a bear in the cave up the trail, or that hostile hunters lurked in the woods. Information was up close and personal; messengers were people who walked a few miles ahead of us. That something important happened in far-off lands, that there even were far-off lands, were probably concepts the Paleolithic man and woman rarely if ever had occasion to contemplate.

It was only starting about seven thousand years ago, with the great empires of antiquity—what we call "civilization"—that we first recognized the need to know what was going on across continents or oceans. Roman emperors, kings of Assyria, pharaohs of Egypt, the great Inca god-kings, and indeed many of their subjects, from traders to diplomats, were concerned with getting information from great distances over lands and seas as fast as possible. In essence, whatever the culture or period, they had one goal: the proverbial annihilation of distance.

The early varieties of technologies for signaling over modest distances included smoke, light, dust, and fire.[3] Each played a role in "foreign" affairs and constituted an early form of delivery of news from afar, if not from abroad. Aeschylus's play *Agamemnon* (written ca. 525–455 BCE)[4] begins with a watchman in the Greek city of Argos waiting to spot "the glow of signal-flame, / The bale-fire bright, and tell its Trojan tale— / Troy town is [taken]."[5] The watchman's long vigil (ten years according to legend) is rewarded when "a beacon-light is seen reddening the distant sky"; in response to the good news of victory in the foreign war, his city, he predicts, will burst into "light, and dance, and song."

For most of history, though, the reporters of distant events walked, ran, rode, or sailed from one point to another, then gave others oral, written, or pictorial messages. When organized, a system of such handing-off of information could be quite successful and helped cement the integrity of vast realms. The Persian Empire of the fourth century BCE, extending from present-day Iran to Lybia and the Ionian Sea, employed an elaborate system of relay-riders and posts. The Greek historian Herodotus tells us, in a phrase that would be adapted thousands of years later to describe the U.S. Postal Service, "There is nothing in the world which travels faster than [they do]. . . . Nothing stops these couriers from covering their allotted stage in the quickest possible time—neither snow, rain, heat, nor darkness."[6] Similarly, the Inca in South America controlled disparate tribes in their Andean empire by use of an elaborate series of beautifully engineered roads with posthouses *(tambos)* every four or five miles.[7] Relays of runners called *chasquis* could span the kingdom quickly. The empires of Assyria, Egypt, Rome, China, and others also built networks of foot, horse, and boat traffic to transmit news. Their swiftness, of course, was relative. One of the last systems used before completion of a transcontinental telegraph line in the United States, the famous Pony Express of the American West in 1860–61, carried messages from St. Joseph, Missouri, to Sacramento (about 1,900 miles on the trails of that time) in ten days.[8]

The fastest message-carrier before Morse's invention was the homing pigeon, used extensively by enterprising journalists in the nineteenth century. These birds (the best came from Belgium) were even thought to be faster on average than a steam locomotive, managing speeds of forty to seventy-five miles an hour. But, as one journalist of the time lamented, they "were not reliable on distances exceeding four or five hundred miles."[9] As this sentiment suggests, messengers, whether feathered or human, had limited utility when it came to foreign news. They were subject to frailty, fallibility, and fickleness, and, in modern terms, they could carry only small amounts of data. The Incas' written communications, for instance, consisted of *quipu,* a cluster of knotted strings that served as mnemonic devices to record numbers.[10] In another example, the modern Olympic marathon race celebrates the legendary feat of a Greek herald who ran 140 miles in 36 hours to report to the council of Athens: "Chairete, nikômen!" (Rejoice, we are victorious).[11] This was hardly a "live from Baghdad" report with color video.

Another constraint of such messengers was that often they were not bringers of news at all but were controlled by the official powers, carrying information for the exclusive use of their warlords or political masters.[12] Only large empires could afford to maintain systems of riders, and when the former fell the latter disappeared as well. Even smaller rider systems were pricey: customers of the Pony Express paid about twenty-four dollars in today's currency to deliver a half-ounce letter.[13]

Yet "open-source" messengers and heralds did exist. News traveled unofficially with merchants and returning soldiers, and the "town crier, spreading news among the local community," was common in the ancient world.[14] There was also a professional class of independent and entrepreneurial "newsmongers" who took on the mantle of the herald and sold information (often unreliable or fantastic).[15] Indeed, because distortions and lies about foreign news were so common, heralds were described as often endowed with powers of "charm, trickery, cunning, and deception."[16] Multiple sources of accurate verification were available neither to emperors nor to common people, and the limitations of "news" affected events in the ancient world. For example, the great Roman emperor Marcus Aurelius's best general, on a distant battlefront, rose in revolt because of a false rumor that the emperor had died. Fancy, it is said, tended to travel faster than fact.

Despite such energy and ingenuity, as historian Paul Valéry once observed of the relatively slow pace of travel during most of human history, "Napoleon moved no faster than Julius Caesar,"[17] and neither did his correspondence. Foreign affairs were affected by such limitations. The Battle of New Orleans, in which Andrew Jackson defeated an invading British army (and thereby established himself as a national hero), was fought on January 8, 1815, although the Treaty of Ghent, ending the War of 1812, had been signed in Belgium on December 24, 1814. Slow communication worked the other way, too, sometimes producing good results. A dispute arose between the government of Abraham Lincoln, lately fighting a rebellion of southern states, and Great Britain over the rights of merchantmen at sea. It was only the *lack* of speed of news, historians have posited, that allowed both governments to develop prudent responses to what each saw as the other's provocations and avoid war.

Such delays—in news arriving and in allowing statesmen and the public to deliberate reactions to that news—became less probable when Samuel F. B. Morse's telegraph accelerated communications. No sooner was a transcontinental telegraph line strung in 1861 than the Pony Express was out of business.[18] Once in motion, innovation came faster and faster. Steam, electricity, and gasoline powered new forms of transportation and high-speed printing presses. Photography and the wireless offered entirely new modes of communication before the end of the century. The pace quickened in the twentieth century, with the rise of radio and television, delivery of news by satellite, and the Internet. By the turn of the millennium, the Industrial Revolution was pronounced dead. We were in the Information Age.

With the steamship and the telegraph, and then other improvements, increasingly commercialized newspapers began to distinguish themselves from their competitors by the proverbial "scoop." Fact-based reporting by professional reporters and editors developed in tandem with technology. Notwithstanding the ongoing debate on the origins of objectivity and the inverted pyramid (the style of reporting that puts the most important information at the top of a story),[19] the fact is that those forms of reporting were well suited for the telegraph.

Technology gave greater geographical reach to journalists but also brought new costs. Press associations, wire services, and syndicates formed to collect information from afar, preventing each newspaper from having to shoulder the often considerable cost of getting the same story. For the first time, a newspaper and its readers had serious expectations about getting up-to-date foreign news. Not long after the transatlantic cable was laid in 1858, this request came to Reuters in London: "PRAY GIVE US SOME NEWS FOR NEW YORK, THEY ARE MAD FOR NEWS."[20]

The first reaction brought on by improved communications technology was optimism about its seemingly self-evident benefits. In 1900, *Scientific American* ran an article on technology celebrating "A Century of Progress in the United States":

> The railroad, the telegraph, and the steam vessel annihilated distance; peoples touched elbows across the seas; and the contagion of thought

> stimulated the ferment of civilization until the whole world broke out into an epidemic of industrial progress. . . . To-day two cents carries a letter to Manila, half way round the world . . . and the New York daily morning papers are distributed in Washington in time to be read at the breakfast table there on the same day of their issue.[21]

Around the same time, the president of the Canadian telegraph system waxed eloquent on the virtues of communication: "The means of communication which are the signs of the highest forms of civilization are the most perfect by aid of electricity simply because they are instantaneous. There is no competition against instantaneousness."[22] The *Electrical Review* went so far as to predict a time "when one can make a tour of Europe without going out of his own house."[23] Guglielmo Marconi's wireless, journalist Ray Stannard Baker enthusiastically observed in 1902, opened a "field of almost illimitable future development."[24]

The celebrations have continued, but voices of doubt have multiplied with each advance. Correspondents of print media have fretted that the ability to file quickly often trumped thoughtfulness. Similar doubts crept in about the virtues of Marconi's wireless and its offspring, radio and television. True, broadcast correspondents gave their audiences vivid sounds and images from abroad, but televisers ignored stories for which they could not get good video. *Los Angeles Times* China hand Robert Elegant saw improved communications technology as a mixed blessing at best. "Marconi ruined foreign correspondence," he said. "I think Morse may have helped."[25]

For their part, policy makers—who followed overseas crises live on television—had more up-to-date information than ever before. Never again would they fight an unnecessary battle the way Jackson did in New Orleans. Yet instant communications created dilemmas that would not have existed when insults traveled no faster than a horseback rider. Officials had to make quick decisions when taking more time would have been better. No president or secretary of state today is afforded the luxury of the weeks of deliberation to craft a response to explosive news from abroad available to Abraham Lincoln and Prince Albert in 1860.

We harnessed satellites, cell phones, and digitization, but controlling the reins is not that easy. Is the new technology helping us to get faster, better, more accurate news? How does the new "instant" foreign correspondence affect the nature of news, the public definition of what is "foreign" and a "correspondent," and the actual carrying out of foreign policy? Unfortunately, most practicing foreign correspondents in their gray years seem to relish what anthropologists call the genre of "tales of decline." In their memoirs, these professionals lament bygone eras of freedom, dignity, adventure, and heroes—no matter that the previous generation of aging foreign correspondents shook their heads in disapproval at that very time and hearkened back to their own, older golden days. In this volume we hope to break the cycle of nostalgia and despair. Technology will not be undone. It is fruitless to pine for the good old days that perhaps never were.

## CHALLENGING BASIC TRUTHS OF FOREIGN CORRESPONDENCE

While there are valid reasons to be concerned about the problems that communications advances bring, this book operates—as we have noted—from a different mind-set, one that is in keeping with the Breaux Symposium out of which it developed. That meeting of minds, conducted each year by the Manship School of Mass Communication's Reilly Center for Media and Public Affairs at Louisiana State University, reconsiders established truths about media. For the symposium that was the basis of this book, we set our authors to assess the accuracy of long-held truisms of foreign news in light of new developments.

In chapter 2, Lucila Vargas of the University of North Carolina, Chapel Hill, and Lisa Paulin of North Carolina Central University ask the basic question, What is "foreign" news? The pat answer is that *foreign news is news about foreign countries.* But is this—or was this ever—so straightforward a definition? Relevantly, the great publishing baron Henry Luce once said that *Time* magazine was written for the "thinking gentleman from Indiana." Self-evident in the 1940s was that such a "gentleman" was a U.S. citizen by birth, heterosexual, and of Anglo-Saxon descent (and probably an Episcopalian). To this ideal reader, naturally, news from Mexico, or China, or South Africa was unambiguously "foreign," as much as news about the Hoosier basketball team's fortunes was "local." Even for the recent immigrant, foreign news in the early twentieth century was often posted as distinct from "American" news. Robert Park, one of the fathers of the Chicago School of Sociology (a group for whom news was a central process in the urban ecology), found that even ethnic newspapers emphasized "Americanism," that is, teaching their immigrant readers distinctive "American" values and ideas (i.e., "new country" vs. "old country" news).

We argue that demography, ideology, and technology challenge the assertion that news can be categorized as "foreign" versus "local." First, more and more Americans are hyphenates uninterested in being blended into a melting pot. Many in the modern non-Anglo majority see America as an extension of their place of birth: they are binational, send money home to their families, and travel back and forth regularly. Many have access through satellites and the Web to "international local" news from their homelands and "local international" news from international sources. One of the editors of this book's first-generation Greek relatives, for example, get their news about Greece from Greek television in Boston, but that is also their source of news *about* Boston, which is extensively covered in the Greek press. Arab Americans may get their news about America from Al Jazeera. Many Mexican Americans read their "local" paper (Spanish-language or English) in Dallas or Los Angeles and find extensive coverage of their homeland to the south. Further, the ethnic press is no longer an agent of assimilation but now of nationalism, of keeping "foreignness" as part of "Americanness." Finally, even non-ethnics can see international news as "local," as many Christian evangelicals may see news about Israel.

The rise of an interdependent global economy and an instant-everywhere communications system has also erased many of the distinctions of what constitutes foreign news for everyone. Is a story about China's petroleum stock-buying spree an international story when, in fact, it directly affects gas prices at our neighborhood station? What about a SARS outbreak in Vietnam that might arrive in our home city by plane within a week? Should a story about the presence of one Osama bin Laden in Afghanistan in September 2001 have been explained to New Yorkers as not of local interest? In an intimately interconnected world, can we afford to use hard-and-fast definitions of foreign versus domestic affairs?

In our third chapter, Steven Livingston of George Washington University challenges the assumption that *people rely on mainstream media for their foreign news.* Certainly, once upon a time we absorbed our knowledge about foreign affairs from Walter Cronkite, the afternoon paper, and Bob the "opinion leader" in our workplace. It would have been logistically difficult, complicated, and time-consuming work for ordinary news consumers to get much more information than these traditional sources offered about, for instance, the second Taiwan Straits crisis of 1958, the Tet offensive, or the Mayaguez incident (and unusual if they bothered). Moreover, such a system made sense because it was assumed that (a) people seek out familiar sources that they can trust for news; (b) only large, traditional news organizations have the structure and funding to cover foreign news exhaustively and accurately; and (c) true "foreign correspondents" are by definition employed only by conventional media organizations.

The new media world (of both technology and economics) overturns these commonplaces in several ways. First, the rise in the 1980s of news divisions that were asked to be "profit centers" for networks has meant substantial cuts in their reporting staffs, operating budgets, and bureau setups in foreign countries. Second, there now exists a tech-enabled marketplace of ideas with many different venues and forms of information about foreign affairs. News sources include not only independent suppliers like "JoeSchmoIraqBlog.com" but also media programming that used to be classified as "entertainment" like Jay Leno and *The Daily Show.* In addition, local news organizations are finding it easier to cover international affairs.

Finally, the "foreign correspondent" exists in various guises. Many corporations (such as Federal Express) and NGOs have their own full-service, in-house news organizations, and other companies (like Bloomberg) provide information for employee and executive decision-making. Indeed, individuals are becoming their own foreign correspondents by creating their personal foreign affairs home pages and resources. Even foreign governments, like that of the Saudis, are building their own news distribution systems to go directly to the American public. If different groups have different needs and uses for and derive different gratifications from news, do we need mainstream media at all? Professor Livingston explores a world in which the channels and sources of information of foreign affairs are almost unlimited.

In chapter 4, Kaye Sweetser Trammell (University of Georgia) and David D. Perlmutter (University of Kansas) dispute another platitude of foreign affairs coverage: *foreign correspondents must be professionals.* Certainly, the modern style of news demands a focus on the people within and behind any foreign affairs story. As the foreign correspondent and travel writer Robert Kaplan observed to one of us, "It is [inconceivable today] to tell a story about what is going on in, say, Egypt without having a vignette about Ahmed the [Nile] fisherman: his life, his worries, and his opinions on foreign policy." Notably, in the past, foreign peoples told their stories through the interpolation (selection, editing, translating) of Western journalists. Only rarely did select foreigners (Madame Chiang Kai-Shek, Winston Churchill, Tokyo Rose) try to influence U.S. foreign policy by speaking directly to large numbers of Americans.

New media technology is radically changing this basic structure of foreign affairs reporting. The advance guard of the revolution is Weblogs, or blogs. These are Internet sites on which anyone—from Harvard professors to thirteen-year-olds—can rant, rave, debate, flame, pontificate, and gush about issues of the day, including foreign affairs. Domestically, a number of these blogs have grown tremendously in readership, including Wonkette, Daily Kos, Little Green Footballs, and Instapundit. The webmaster of the latter recently noted that hundreds of thousands of people check his site daily for information about domestic and international affairs and that his "correspondents" from all over the world, including Iraq and Afghanistan, send him reports and pictures and thus "cover" the news. Blogs like these have gained power as well by their attacks on mainstream media; conversely, many in mainstream media consider blogs a threat to the basic paradigm of journalism as an industry.

While a number of writers in this book will refer to blogs that, in part, report and comment on foreign affairs issues, the goal of this chapter is to understand one particular novel development in the blog phenomenon. For the first time in history, people who are part of a foreign affairs story—but not in the capacity of enemy warlord, propaganda minister, or other kind of powerful elite—are able to speak directly to the American people and possibly influence foreign affairs. What is happening and what could happen when Ahmed the fisherman has his own public affairs Web site? Some foreign blogs, like that of Iraqi Salam Pax, have become famous. Beyond individuals, groups—including those in opposition to America such as Al Qaeda—have created their own Web sites to bypass the editing (and presumably ethical constraints) of mainstream American media. (Their pictures *are* out there in the news stream; they often appear in Arab media, for example, as uncredited stills or videos.) Independent foreign bloggers, writing in English with some expectation of an American audience, are exerting an increasing influence on the coverage, creation, shaping, and operating of foreign affairs; they are both part of the story and tellers of the story of news from abroad.

The fifth chapter, written by Louisiana State University's Margaret H. DeFleur, explores a factor in defining "foreign news" that challenges a paradigm held by many, especially critics of journalism, ranging from academics to leaders in the developing world. It has long been held that America has engaged in a form of "mediated" imperialism, that is, *as American media dominate the world, American cultural and political hegemony follow.* By *media* we do not here mean the CBS Evening News or even CNN International but entertainment media. We feel consideration of such media in this book is warranted because, although there was a time when it was clear where "hard" (domestic and foreign) news ended and entertainment began, today that distinction is blurred in the media content and news produced in the United States and disseminated nationally and globally. Such a development has important implications, not only from the perspective of American audiences but also for the future of foreign affairs.

Since 9/11, much of our foreign affairs reporting has been concerned with "why they hate us," or variations on the questions of how Americans are viewed abroad and why foreigners often hold negative opinions and attitudes toward the United States and/or its people, leaders, and policies. The role of cultural media—the intoxicating and not always salubrious fare served up by Hollywood movies, television programs, and video games—has been little examined in this debate. In fact, a standard assumption has been that American entertainment media projected by various technologies around the world has enhanced U.S. political power. After all, if they are watching *The Sopranos, Rambo,* and MTV, listening to Madonna and Jay-Z, and playing Halo 2 and Sim City, aren't foreigners being co-opted into admiring Americans and their society as well?

There are problems with that assumption, argues Professor DeFleur, based on extensive research on the sources of anti-American attitudes abroad. First, it is a mistake to ignore the part played by popular culture as a component of foreign affairs or to assume that cultural exports buy favorable attitudes or political power in the "street" of the Arab—as well as the European, Asian, African, Central and Latin American, and even Canadian—world. The opposite may be the case. Exporting mass media entertainment products—that is, various forms of popular culture produced in the United States—to worldwide media markets where they are eagerly received appears to have serious and negative consequences for American foreign affairs.

Indeed, "media imperialism"—our cultural domination of the world—has in fact a powerful blowback effect in foreign affairs. Among the majority of the foreign youths studied, ordinary Americans are assumed to be violent and often engaging in criminal activities. American women, as depicted in media entertainment, are thought to be sexually immoral—especially when judged against local standards for female conduct. Few of those studied had ever met an American or traveled to the United States. Their principal source of knowledge about the people who live in the United States was what they saw depicted in movies, TV programming, and even news. Given the negative images that these sources frequently present, it is not surprising that foreigners have learned to judge Americans harshly. After all, they have "seen with their own eyes" what Americans are actually like. In sum, as Professor DeFleur argues, if they are to understand hostilities toward American citizens and their society, political leaders, journalists, policy makers, scholars, and commentators must assess U.S. entertainment media exports as a major component of foreign affairs.

Chapter 6 examines a long-held assumption of the news business itself, that *the greater the autonomy of foreign correspondents, the more it allows them to create original and locally flavored stories.* John Yemma of the *Boston Globe* argues that as recently as a decade ago, foreign correspondents were the princes of the profession because, in part, the home office did not closely supervise them. They roamed the world subsidized by expense accounts, looking for interesting stories, dropping into famous hotels near war zones, and filing dispatches for a home audience of editors and readers who relied on and enjoyed their reports. Their main means of staying connected with headquarters was the telex machine at their hotel. Overseas phone calls were expensive and consequently rare, especially for journalists working for smaller newspapers. Editors would direct foreign correspondents through terse telexes such as "Much like your latest exBeirut. Interested in piece on Kurdish sitch. Allbest." As a result, for the most part a correspondent would be left to develop a story on his or her own, often after reaching a consensus by swapping notes with the other correspondents (the "tong" or pack) at the breakfast table. With little up-front consultation with the editor, a reporter would simply send a heads-up message ("Filing 800-wds by noon on Syrian efforts to derail peace talks") followed by the full story.

In sum, foreign correspondents enjoyed extraordinary autonomy. Their prose was edited lightly, in part because the edits were not played back to the correspondent. If a substantial rewrite were needed or one story had to be combined with another, that was the job of the foreign desk. The correspondent essentially filed and moved on. That process allowed him or her large amounts of time to travel, interview, and pursue local color. Often this made for vivid dispatches, but with no one directly supervising or covering the same beat, the stories filed by "our man in Damascus" might be factually suspect as much as they were romantically colorful.

The glory days of the "independent" foreign correspondent are gone. Technological advances, from cell phones to the Internet, have radically changed the paradigm. Today's foreign correspondent is interconnected. Except for time-zone differences, the information loop with a journalist in Baghdad or Beijing is almost identical to that of a coworker on the city desk or Capitol Hill. Even in war zones or in remote areas like eastern Congo, correspondents stay in touch via international cell phones and e-mail. This situation allows extensive conversation between foreign editors and correspondents before a story is commissioned. It facilitates correspondents' read-backs of their stories in the editing process and allows them to see instant reaction to their stories once published. The Internet and various subscription databases like LexisNexis and Factiva provide a traveling reporter the same global library that anyone at home can use.

These new technologies drastically decrease the autonomy of foreign correspondents and ensure that editing is more cooperative, transparent, and iterative. A correspondent can no longer complain that Boston or Chicago or LA took liberties with his or her prose when the correspondent is at the other end of an "IM" exchange throughout the production cycle. This short leash makes a foreign correspondent's job less colorful (although there is still plenty of local color to be had) and means that editors, readers, and media critics scrutinize her or his dispatch more closely. The feedback loop is almost instantaneous. Because of this, foreign correspondents have become better at sourcing—pushing for named sources—and are less likely to take liberties with descriptions and quotes than in the past. Autonomy, Yemma concludes, did not (and does not) breed better foreign correspondence; the "managed" foreign affairs reporter may, in fact, produce better and more reliable work.

Chapter 7, written by Emily Erickson and John Maxwell Hamilton, both at Louisiana State University, looks at a similar old saw of foreign correspondence: *foreign news should be covered by professional journalists posted permanently overseas and not parachute drop-ins.* The ideal foreign correspondent, common wisdom once held, was the longtime reporter who spent years establishing contacts and sources; learning the culture, politics, and language of foreign lands; and acquiring the general facility to get around in any country where their editors sent them. Those who had no foreign experience and no intention of staying around to get it were disdained, as in the case of lead anchormen popping into big developing foreign stories. Its critics viewed parachute journalism as a device with which to avoid the cost of posting permanent correspondents abroad. As this argument went, foreign correspondents were a breed apart that had little in common with the stay-at-home types.

Although they insist it is imperative to have permanent foreign correspondents based overseas, Hamilton and Erickson argue that technological advances in our ability to travel far and fast relatively cheaply offer positive models to supplement that traditional approach. For medium- and small-sized media in local markets, the costs of travel are reduced to the point where they can realistically consider sending reporters abroad. Furthermore, with world cultures converging, is Seoul or Athens as strange to a parachute journalist as it would have been, say, in 1900? There is also much more information available to the parachuting journalist. Elaborate local networks of stringers and fixers can facilitate drop-ins. We might also question whether the "old hands" were as expert as we romanticize them to be. On the one hand, could someone be an expert on "Asia"—Singapore to Burma, for example—where they were likely to be dispatched if a coup broke out? On the other hand, can someone with a specialization in a subject area, rather than a geographic area, do a better job on certain complex stories? The "old hands" were often criticized for building up encrusted biases and tending to go repeatedly to the same sources, often government staffers or vocal locals, for their stories, so might a naïve newcomer bring some fresh insight?

Finally, one of the problems with traditional foreign correspondence has been its lack of resonance with the great body of Americans except in times of palpable crisis. Parachutists are particularly well equipped to bridge this gap, for they are rooted in their communities. Resonance with readers, viewers, and listeners is especially critical now. Global interdependence is having an enormous everyday impact on Americans' lives, but the public often doesn't see the connections clearly.

Thereafter we proceed to question even more fundamental assumptions of foreign news—or of any news. Philip Seib of the University of Southern California challenges the professional standard that *for foreign news, faster is better.* We know that the idea of getting there first with the big story is an old news value, going back to the era of competing yellow sheets sending the newsboys into the street crying, "Read all about it! Special edition!" If anything, new technology has ramped up the competition for recency. Today, cable news directors will boast that they trumped the competition for a breaking story by a few minutes. In an age of nukes and terrorism, we need news fast to contend with rapid developments. It would seem, then, that this "truth" is a case of new media tech fitting well with a time-honored journalistic value. We get our foreign news faster than ever, and this seems objectively to our advantage.

Professor Seib critiques the notion that "faster is better" on several levels. First, in the pre-satellite days, journalists had a "news cycle" in which to consider whether news was news and was fit to print. Second, the modern "first draft of history" is often full of errors. There is a lot of evidence that coming out too quickly with a story, especially murky information from a foreign land based on questionable sources and sketchy rumors, shouldn't be a news value at all. Are there ways that we can use new technology to become more accurate, more reliable, more authoritative, rather than just faster? We now have context stripped in favor of action. Slowing down may improve the quality of the story and increase public understanding. Furthermore, does not the worship of speed lead to a dearth of style and wit? Finally, do we know that people actually want scoops? Does it matter to them who is first?

We close our study by asking former U.S. State Department under secretary for management Richard Moose to step back and give us the policy makers' perspectives—that is, to survey what we have learned about the relations of new media technology, economics, styles and processes, and foreign affairs coverage and issues. His afterword provides an insight by and for the policy makers, those people in government actually charged with conceiving and shaping foreign policy and putting programs into effect. How does the new media world affect what they do? Has this changed over time? Is policy driven by "media" or is it "sold" in anticipation of media coverage? Do new technologies of communication affect the way policy makers learn about issues and disseminate their own information on them? In all, how does communication technology affect policy making in the realm of foreign affairs? Moose's conclusion is reassuring. We may live in an Information Age that shapes the execution of American foreign policy, but in the end what counts the most is the policy itself.

In his play *Endgame,* one of Samuel Beckett's characters comments, "Ah, the old questions, the old answers, there's nothing like them!" For most of human history the technology of foreign news gathering was static. No more. Not all the questions and answers have changed, but many in this instant-messaging environment have. Because foreign news is so important to effective policy making and citizenship, we imperil ourselves by failing to understand the changes technology brings and how we can wring the best practice out of those changes. Whatever errors the authors of this book may make in their judgments, we hope it usefully tests some of the old satisfying answers that may lead us astray.

NOTES

1. Mike Wendland, "From ENG to SNG: TV Technology for Covering the Conflict with Iraq," Poynteronline Weblog, March 6, 2003. http://www.poynter.org/content/content_view.asp?id=23585.

2. Quoted in John Keane, *Tom Paine: A Political Life* (1995; repr., New York: Grove, 2003), 231.

3. Garrick Mallery, *Sign Language among North American Indians, Compared with That among Other Peoples and Deaf Mutes* (1881; repr., The Hague: Mouton, 1972), 300; Sue Northey, *The American Indian* (1939; repr., San Antonio: Naylor, 1954), 35; P. Southern, *Signals versus Illumination on Roman Frontiers* (London: Britannia, 1990), 233.

4. G. Wilson, *The Old Telegraphs* (London: Phillimore, 1976), 1–4. Aeschylus's play *Agamemnon* alleges that the Greeks sent home from Troy a signal of victory, so beacons must have been in use for long-distance communications by at least 525–455 BCE.

5. Aeschylus, *Agamemnon* (ca. 525–455 BCE). http://classics.mit.edu/Aeschylus/agamemnon.html.

6. Sian Lewis, *News and Society in the Greek Polis* (Chapel Hill: University of North Carolina Press, 1996), 60; Herodotus, *The Histories,* trans. Aubrey de Selincourt, rev. by A. R. Burn (New York: Penguin, 1972), 556.

7. John Hemming, *The Conquest of the Incas* (New York: Harcourt, Brace, Jovanovich, 1970), 101–2.

8. Tom Logsdon, *Mobile Communication Satellites* (New York: McGraw-Hill, 1995).

9. Frederic Hudson, *Journalism in the United States, from 1690 to 1872* (New York: Harper and Brothers, 1873), 597.

10. Hemming, 61, 553.

11. Frank J. Frost, "The Dubious Origin of the 'Marathon,'" *American Journal of Ancient History* 4 (1979): 159–63.

12. Lewis, 60.

13. Philip L. Fradkin, *Stagecoach: Wells Fargo and the American West* (New York: Simon and Schuster, 2002), 29–30.

14. Rosalind Thomas, *Oral Tradition and Written Record in Classical Athens* (Cambridge: Cambridge University Press, 1989), 238.

15. Lewis, 75.

16. Linda S. Frey and Marsha L. Frey, *The History of Diplomatic Immunity* (Columbus: Ohio State University Press, 1999), 14–15.

17. Fernand Braudel, *Civilization and Capitalism, 15th–18th Century,* vol. 1, *The Structures of Everyday Life: The Limits of the Possible* (New York: Harper and Row, 1981), 429.

18. Tom Standage, *The Victorian Internet: The Remarkable Story of the Telegraph and the Nineteenth Century's On-Line Pioneers* (New York: Berkley Books, 1999), 59.

19. For examples of this debate, see David T. Z. Mindich, *Just the Facts: How "Objectivity" Came to Define American Journalism* (New York: New York University Press, 1998) and Michael Schudson, *Discovering the News: A Social History of American Newspapers* (New York : Basic Books, 1978).

20. Standage, 153.

21. Edward W. Byrn, "A Century of Progress in the United States," *Scientific American,* December 1900, 402–3.

22. Quoted in Carolyn Marvin, *When Old Technologies Were New: Thinking about Electric Communication in the Late Nineteenth Century* (New York: Oxford University Press, 1988), 193.

23. Ibid., 200.

24. Ray Stannard Baker, "Marconi's Achievement," *McClure's,* February 1902, 296.

25. John Hohenberg, "The New Foreign Correspondence," *Saturday Review,* January 11, 1969, 115.

# 2

# RETHINKING "FOREIGN NEWS" FROM A TRANSNATIONAL PERSPECTIVE

LUCILA VARGAS AND LISA PAULIN

Under globalization, what is foreign news? The notion of "foreign news" implies that internal and external events are neatly separated, yet what characterizes globalization is precisely the opposite. Globalization has been defined as the integration of the world economy, the mixing of cultural traits, the increasing significance of political organizations and processes that bond nations, and, in general, as the transnational connections of various kinds that new information and communication technologies have made possible. The idea of "foreign news" also presumes a sharp difference among residents of different nations. One of the traits of globalization, however, is the mobility of people across national borders. Moreover, the idea of "foreign news" implies that events that affect "them" do not have a direct impact on "us." This is, of course, a highly questionable assumption.

In this chapter, we explore how the transnational practices that came with globalization have cast doubt onto the very notion of "foreign" news. The integration of the global economy has created strong links among regions belonging to different nations. The politics of many countries have spilled over their national borders, and their new political maps incorporate residents of other countries. Many people, immigrants in particular, have a passionate interest in other countries' local politics, and they are directly addressed by politicians and news media from abroad. While some have dual or multiple citizenship, others do not, but their sense of belonging transgresses conventional forms of citizenship. Haitians living in the United States, for example, are often called "Haiti's tenth province," a term popularized by Haiti's president Jean-Bertrand Aristide. The Mexican Congress recently passed a bill granting Mexican citizens who reside in other countries the right to vote in presidential elections. And in places like Washington Heights, New York, immigrants from the Dominican Republic participate in both the U.S. Democratic Party and parties of the country they left behind. As it is for many Haitians, Mexicans, and Dominicans who live in the United States, for millions of people throughout the world news about their home countries is not foreign but near and dear.

Transnationalism "broadly refers to multiple ties and interactions linking people or institutions across the borders of nation-states."[1] It is a new conceptual model that allows one to think in terms of flows and networks. We argue that the category "foreign/international news" is inadequate to capture many of today's economic, political, cultural, and social interconnections that seem to overlook political borders. We suggest the category "transnational news" to group stories about events and processes that happen in social spaces that traverse national borders.

Our interest lies in the ways in which the rapid diffusion of new information and communication technologies has complicated customary distinctions between "foreign" and "domestic" news. These technologies have dramatically changed the gathering, production, and distribution of foreign news, but as we discuss below, they have also been critical in the development of new publics and new subjects or "beats" for what can be called "transnational affairs reporting."

The notion of the Jewish diaspora has been fruitfully applied to other migrant groups who retain heartfelt attachments to the homeland. We find this notion helpful in sorting out what international news means for transnational ethnic groups. Diasporic people develop bicultural identities, or a sense of belonging to more than one setting. For many, information about the homeland is as relevant and familiar as news about their present place of residence. Often living in the midst of economic, cultural, and political processes that span nations, diasporic populations rely on transnational social networks. One of the most insightful commentators on the use of new media by diasporas, scholar Karim H. Karim, notes that in the last decade these networks have been greatly enhanced by the "extensive use of on-line services like the Internet, Usenet, Listserve, and the World Wide Web."[2] At the same time, Karim points out that digital broadcasting satellites and digital compression have facilitated the exponential growth of diasporic programming. "Ethnic media," says Karim, "have frequently been at the leading edge of technology adoption due to the particular challenges they face in reaching their audiences. The relatively small and widely scattered nature of communities they serve has encouraged them to seek out the most efficient and cost-effective means of communications."[3]

Other powerful transnational dwellers are the global elites, which include the managerial class of transnational corporations (TNCs), government officials, and others such as sojourners and long-term tourists. Yet the vast majority of transnationals are political exiles, refugees, and, especially, guest workers. International migrants—most moving from southern to northern countries—have become huge media publics as well as significant actors in international affairs. The numbers of transnational dwellers are staggering and they are expected to rise. In the year 2000, an estimated 175 million people were living outside their country of birth.[4]

Although we talk about various transnational spaces, to illustrate with more detail what we mean by "transnational news" we focus on U.S. Latinos/Hispanics[5] because they engage in intense transnationalism and they are representative of the migratory networks created by the new international division of labor. Scholars Marcelo M. Suárez-Orozco and Mariela M. Páez remark that "even as Latinos enmesh themselves in the social, economic, and political life of their new lands . . . they remain powerful protagonists in the economic, political, and cultural spheres in the countries they left behind. Latinos are emerging as 'hemispheric citizens.' Latino remittances and investments have become vital to the economies of varied countries of emigration, such as the Dominican Republic, Mexico, and El Salvador."[6]

## FOREIGN, INTERNATIONAL, OR TRANSNATIONAL NEWS?

It seems odd that the adjective *foreign* became associated with stories about events occurring beyond national borders but having either cultural, geographical, or functional proximity to the audience—characteristics that have been major determinants of what gets covered in foreign news. The first global news network, CNN, rejected the traditional adjective and popularized a less loaded term, *international.* Foreseeing satellite technologies' possibilities for global interconnectivity, Ted Turner launched CNN in 1980, and aware of the contradiction between interconnectivity and lack of familiarity, he made it clear that the network's policy was to avoid the word *foreign* because it implied "something unfamiliar."[7]

Although the notion of "foreign/international news" has become quite hazy, the traditional view is that it refers to "hard" news about political and economic events that happen abroad. Veteran journalist and current editor of *Foreign Affairs* James F. Hoge, for example, says that the coverage of the death of Princess Diana was not foreign news. "The everyday stuff of foreign news," claims Hoge, "consists of political and economic events that raise policy issues and force governments and people to choose."[8] As scholar Margaret DeFleur discusses in chapter 5, the sharp distinction between news and entertainment has eroded, yet Hoge's view still represents the conventional wisdom.

"Foreign news" is a category used by news producers to choose and place stories in predetermined locations in their publications and broadcasts. So what counts as this type of news depends on what is in news producers' heads—rather than on audiences' minds. A Pew survey asked 218 editors responsible for deciding what stories are published in large U.S. newspapers (circulation 30,000 plus) to define "international news." The findings: "[S]even out of ten editors defined the term with phrases such as 'events outside the United States,' 'news happening outside our borders,' 'breaking news from across the globe,' and 'news from any country other than the United States.'" Only one in six editors came close to what we call transnational news: they defined this news as "international stories that have some effect on the United States" and "an event or issue that is occurring somewhere else but has local impact."[9]

Scholar Steven B. Crofts Wiley astutely argues that in communication research the nation "serves as the fundamental point of reference against which other structures and processes are defined" as either local, international, or global.[10] Extending his point to news, one can see that the nation-state has been the point of reference in defining foreign news. In the last few years, however, new technologies have enabled ordinary citizens and relatively small news organizations like Al Jazeera to gather, produce, and distribute news to mass audiences that span several nations. The geography of these audiences is not based on national borders but rather on boundaries related to ethnicity (e.g., the 30 million Kurds targeted by MED-TV), affinity (e.g., the 11 million members of the Ahmadiya sect targeted by Muslim Television Ahmadiya International), or interest (e.g., finance, global warming, or human rights).

Globalization has de facto blurred the lines among many nations and the conventional distinction between "foreign" and "local" news. Former editor of the *Denver Post* Glen Guzzo argues that because "the definition of local news has changed," regional newspapers have to cover international stories. "As many stories in American cities become inextricably bound to events in Afghanistan, Iraq or Mexico," Guzzo says, "local news becomes more difficult to master and more expensive to gather. But without the commitment to do it right, local news becomes shallower—and more expendable to readers."[11] To us, local stories "inextricably bound" to places beyond the national border may be more properly called transnational.

Other commentators think that what has changed is the nature of foreign news. Media columnist William Powers says that "foreign news is out there in great profusion these days, particularly online, but it is a different kind of foreign news. While the old foreign news had an air of urgency that was a product of the cold war and technological constraints, this new foreign news is diffuse, many-layered, sprawling, chaotic, and terribly complicated . . . like the world itself."[12]

Innovative news-making practices in the mainstream media show that a unique, different type of news has emerged. The most telling example is the *Cincinnati Enquirer*'s page called "Crossing Borders," but there are many other efforts to inform ordinary citizens of the relevance of transnational events for their daily life. The American Society of Newspaper Editors Web site has a document titled "Bringing the World Home" with links to stories that give an idea of the best practice of what may be called transnational news making.[13] As we discuss below, media catering to immigrants has always given special attention to transnational news. It is typical of immigrant media, for example, to treat news about their audiences' countries of origin—which is transnational news for their audiences—differently from news about the rest of the world.

## MARKET GEOGRAPHIES

Long ago, marketers recognized enduring affiliations based not on nationality but rather on linguistic or other affinities. They segmented populations into markets, drawing borders according to real and imaginary cultural distinctions and similarities. Such is the case of the ethnic markets, which do not correspond to the geographic boundaries of nation-states. Marketers also devised markets in line with other distinctions, such as age (e.g., the youth market) and religion (e.g., the Christian evangelical market).

To a great extent, transnational media corporations have replaced nations with markets, creating a geography that guides their production and distribution of films, television programs, music recordings, etc. Jon B. Alterman, Middle East Program Officer in the U.S. Institute of Peace's Research and Studies Program, argues that transnational media are creating transnational regional communities in both the Arab world and Latin America. "Transnational media," he writes, "are doing what statesmen and warriors have been unable to do. Building on common language and common heritage, the people of both regions are beginning to come together in ways that would have seemed wildly utopian only ten years ago."[14]

Such market geography, however, assumes that information is merely a commodity like sugar or bicycles, disregarding the public service ideals cherished by international correspondents. While the segmentation of the audience works well for marketers' purposes, it presents grave problems for the functioning of democracy. For example, under the marketer's logic, stories about the Asian economic crisis would target mostly the Asian American market. This strategy would leave other Americans in the dark regarding the local impact of the crisis, and therefore incapable of developing informed opinions about foreign policy. Richard Read of the *Oregonian* showed that foreign stories can be told in a way that is interesting to many readers. He won a Pulitzer Prize for a story that explained the local impact of the Asian economic crisis by following the flow of Oregon's frozen french fries to Singapore.

## LOOKING AT THE BIG PICTURE

"Foreign News Shrinks in Era of Globalization," stated the headline of a *Los Angeles Times* article published in the aftermath of 9/11.[15] The headline sums up a deep anxiety that has prevailed in U.S. journalistic circles since the late 1990s. It has often been asserted that the major reason for the shrinking of the international news hole since the early 1990s was the end of the cold war. While this may be true, such shrinking also signals a more profound transformation in the transfer of information across national borders.

At the same time that the foreign news hole became smaller and "softer," other developments with far-reaching consequences for the coverage of foreign affairs were occurring in the media landscape. One was the tremendous growth of multinational business and financial news services, especially Bloomberg and Reuters. A second development was the emergence of transborder communication networks based on affinity or interest. A third development was the expansion of media catering to populations that are geographically dispersed but unified along linguistic or ethnic lines; the most notorious example of these media is Al Jazeera television. These three developments, of course, could not have occurred without satellites, the World Wide Web, and digital technologies, but their diffusion did not happen in a political and economic vacuum.

To appreciate the roots, and possible directions, of today's exchanges of information across national borders, one needs to consider the broad political-economic changes associated with globalization. In a thoughtful essay about the questions raised by globalization for democracy, political scientist James Anderson offers a number of insights that help us to make sense of the current transformation of foreign news. As Anderson notes, the geographic diffusion of capitalism and the rise of the liberal democratic state came with profound changes. Important for our purposes are the economic liberalization and deregulation of the information and communication sectors. They made possible the current concentration and conglomeration of the media industry, as well as the global drive toward privatization of media and telecommunications. Paradoxically, while this trend advanced the ideals of the free press across the globe, it also resulted in the decline of the public service ideals in news organizations in the United States.

Media insiders agree that changes in the economics of the media industry, its for-profit orientation, and its dependence on advertisers are crucial reasons behind the decline in foreign news. Leonard Downie, executive editor of the *Washington Post,* says that his paper and similar dailies "are committed to national and foreign coverage and will remain so. The big question is what the large ownership will do—the networks and the chains—and I'm skeptical they will change. They put foreign news at the bottom of their priorities. They thought it turned audiences off and drove readers away. Will they now put public service ahead of profits?"[16]

New technologies and dot-com start-ups played a part in the emergence of a highly competitive environment where there is little room for public service ideals. Public broadcasters all over the world have been under growing pressure to adopt new digital technologies and to become profitable. In a remarkable document that defends the public service ideals of the BBC, Chairman Michael Grade says that the BBC needs to renew itself and "to deliver those ideals in a digital world." But then he warns about what he sees as the potential perils of that world, stating that it "contains the possibility of broadcasting reduced to just another commodity, with profitability the sole measure of worth. A renewed BBC, placing the public interest before all else, will counterbalance that market-driven drift towards programme-making as a commodity."[17] It is important to note how Chairman Grade conflates technology and ideology—that is to say, the "digital world" with the market-driven media model. His words imply that the diffusion of digital technologies has occurred within a political-economic environment that looks right through the ideals of public service.

But Richard Tait, editor in chief of Britain's Independent Television News (ITN), thinks that British television has maintained its commitment to foreign affairs reporting during a time of drastic cuts. "The secret," he says, "has been greater efficiency—multiskilling and the introduction of cheaper, more flexible, digital equipment." For Tait, the change has been in style rather than substance. "What has changed is the range of styles of news and the ways in which news broadcasters can connect with their viewers, listeners and multimedia users." He goes on to explain how ITN produces different newscasts for different audiences, including "S News—a service which still covers the mainstream agenda, but in a style which appeals to a younger audience."[18]

While changes in the style of foreign news have responded to marketing and economic needs, they are also linked to the use of digital technologies. Writing for the *American Journalism Review,* Lucinda Fleeson observes that "a digicam revolution has created a breed of correspondents who travel light, often working alone, producing intimate you-are-there reports for a fraction of the cost of sending a traditional crew. The new technology offers a promise of faster, less produced, more informal stories that not only could increase the amount of foreign news on television, but inject new style."[19] Yet Fleeson notes that the promise has not been fulfilled in commercial television: "Some of the most penetrating and analytic foreign reporting on the air is found in documentaries that run on the financially strapped Public Broadcasting Service, often in the work of young videojournalists, who sometimes finance their own projects or rely on philanthropic grants."[20]

## NEW ACTORS IN INTERNATIONAL RELATIONS AND NEW BEATS FOR FOREIGN AFFAIRS REPORTING

Like many other commentators on the changes brought about by global market forces, Anderson says that these forces have diminished the power of states to set priorities and policies and that many decisions that used to be made by governments (and presumably by the people in representative democracies) are now made by private corporations, or by supra-state political institutions. The erosion of state sovereignty has significant consequences for foreign affairs reporting. Since a great deal of international news has been about the activities of state officials, it follows that governments with diminished power are less newsworthy than they used to be. State officials are no longer the only—and oftentimes not even the most important—actors in the political and economic events that have been the staple of foreign news.

In particular, TNCs and the managerial elite have become dominant actors in foreign news. In an article about the "nouveau foreign correspondents," Joe Strupp writes: "After decades of focusing on the U.S.-Soviet power battle, political upheavals from Iran to El Salvador, and the threat of nuclear war, international news today routinely leads with the latest financial market swings, European currency fluctuations, and the newest merger of a U.S. corporation with an Asian partner."[21] A brief look at the Foreign Press Association of New York's Web site proves his point. The site's page titled "Writing about Everything That Is Worth Some News" contains nine stories; except for one story about soldiers missing in action in Vietnam, all are business stories. They include one interview with the CEO of General Electric, another with the CEO of Merrill Lynch Asset Management, and another with the founder of Mesa Petroleum, which is the largest independent producer of gas and oil in the United States.[22] In their coverage of big business, the mainstream media have reflected the shift toward transnational governance.

The mainstream media's new focus on TNCs and the managerial elite, however, has not been accompanied by an equivalent attention to other significant actors shaping transnational and global policies today. Anderson argues that globalization brought with it a "spectacular increase in transnational governance"[23] that is now implemented by global and regional nonstate actors. In addition to TNCs, he lists the "Bretton Woods trio" (the World Bank, the International Monetary Fund, and the World Trade Organization), supra-state regional trading blocks, transnational associations of regions, nongovernmental organizations, and transnational social movements. All these institutions act in transnational spaces, and the impact of their actions is felt in several countries. Reporting about these institutions may fall better in the category of transnational news than in the conventional category of foreign news. The mainstream media's insufficient attention to, or even disregard of, nonstate actors such as nongovernmental organizations and transnational social movements raises disturbing questions. By neglecting to report the significance of new transnational actors for international relations, are the media sketching a picture of the global situation that fails to match its current complexity? Most importantly, what are the consequences of such media representation for the formation of public opinion and, consequently, for the making of U.S. foreign policy? These urgent questions demand that "foreign" affairs reporting be grounded on a model of international affairs that goes beyond the traditional model, which is based on the idea of the nation-state. Transnationalism offers the conceptual underpinnings for a model of "foreign" affairs reporting that, without leaving aside the major role played by the state, foregrounds the transborder flows of people, ideas, capital, and information that characterize the networked global society.

It is crucial to underline that new media technology now allows other players—such as transnationals—to essentially revise and extend, even contradict the scripts of mainstream media about international events. For example, in January 2006 it was reported in many major American newspapers and network television programs that several people had died of the so-called bird flu in Turkey. But in each case the stories failed to note that the initial dead, from a family in southeastern Turkey, were ethnic Kurds, a people long persecuted by the Turkish government, who probably could not even read the Turkish-language health warnings in the region.[24] Only Kurdish bloggers and others belonging to advocacy networks concerned with Kurdish issues offered the missing political angle to the story.[25] Although blogging can clarify and extend conversations surrounding events, many questions remain about its influence and potential to impact mainstream media messages.

## TRANSNATIONALISM FROM BELOW

Transnationalism is a slippery term often used with various meanings. In an elaborate analysis of the uses of the term, anthropologist Stephen Vertovec mentions two major themes that serve to develop the idea of transnational news. The first is that of a "transnational community," which is made up of networks of social relationships that span geographical distances, and the second is the idea of transnationalism as a "site of political engagement."

Through the lens of a traditional economist, the major transnational networks are those of TNCs. "TNCs represent globe-spanning structures or networks that are presumed to have largely jettisoned their national origins. Their systems of supply, production, marketing, investment, information transfer and management often create the paths along which much of the world's transnational activities flow," says Vertovec.[26] In addition to raw data and information directly related to a corporation's business, intracorporation services supply foreign affairs news to employees. Scholar and former foreign correspondent John Maxwell Hamilton and Eric Jenner, former editor of the *New York Times'* Web page, argue that "in-house news and information gathering" is one of the new kinds of foreign correspondence. "Virtually every global corporation these days has a computer-linked network in which 'staff reporters' provide original information as well as news summaries to employees around the world."[27] Either employed by or closely linked to TNCs is the transnational elite. TNCs and the transnational elite are both actors in and consumers of one of the most vibrant types of foreign correspondence today: business journalism. A considerable amount of the transnational news that circulates around the world is business news. The business of gathering, packaging, and distributing this news is highly profitable, and—no doubt—those who are profiting are members of the transnational prosperous minority.

But unorthodox views of the economics of information production point to the growing importance of news produced by individuals motivated by reasons other than financial profit. Law scholar Yochai Benkler argues that "in the past decade and a half we have begun to see a radical change in the organization of information production. Enabled by technological change, we are beginning to see a series of economic, social, and cultural adaptations that make possible a radical transformation of how we make the information environment we occupy."[28] In his view, such "radical transformation" has been set in motion by emerging patterns of information production and exchange that are challenging the industrial, capital-intensive, and centralized patterns that have prevailed over the last 150 years. The old patterns comprise both market and nonmarket versions. In the market patterns followed by TNCs, news is viewed as a commodity, and behavior is regulated by market and proprietary principles; in the nonmarket patterns followed by states, news is viewed as a social good, but its production is also industrial and highly centralized. Benkler says that the patterns of information production made possible by the microelectronics revolution are nonproprietary, nonindustrial, and "radically decentralized." Economic behavior in what he calls "the networked information economy" is regulated by the principles of the "commons." In "the commons mode of production" information, knowledge, communication, and culture are produced and exchanged through social practices such as sharing and peer-production. Information-sharing through online networks is just one of the numerous cases of social production that Benkler argues are emerging as alternatives to markets and firms. His social theory of the Internet analyzes many innovative practices, including free software, municipal broadband initiatives, and peer-production projects like Wikipedia.

Benkler's theory helps, first, to situate the analysis of "foreign" news within a broad framework of profound social, political, and cultural change; second, it helps to explain the tremendous growth of transnational civil society networks in the last decades. The allusion in the title of his book *The Wealth of Networks* to Adam Smith's influential *The Wealth of Nations* suggests the huge challenge that social production poses to capitalism and the liberal democratic state as we know it. The "basic claim" of his book, he says, "is that the diversity of ways of organizing information production and use opens a range of possibilities for pursuing the core political values of liberal societies—individual freedom, a more genuinely participatory political system, a critical culture, and social justice."[29] Collective struggles for the pursuit of these values are occurring at all levels, from the local to the global, but they often occur transversely, that is to say, through transnational networks of people who share a common concern.

In this sense, transnationalism can be seen as a "site of political engagement," as Vertovec argues. According to him, substantial political activity in the global public sphere is carried out by two types of network organizations: the international nongovernmental organization (INGO) and the transnational social movement organization, such as the indigenous peoples movement. Other political scientists, like Margaret E. Keck and Kathryn Sikkink, do not distinguish between the two and see both as "transnational advocacy networks" (TANs). Although there were networks of people advocating across borders against issues such as slavery or foot binding in China well before the advent of the Internet, the growth of TANs in the last few years has been astounding, and their communication methods have become extremely sophisticated. The amorphous and transient nature of many coalitions of small networks makes it difficult to estimate the number of TANs. Nonetheless, since many TANs are linked to INGOs, the fact that in the year 2002 there were 282,851 INGOs in the world gives an idea of the proliferation of TANs in the last few years.[30]

Through TANs, individuals from many countries come together to carry out political work aiming to change the status quo. The overarching issues—the environment, the status of women, and human rights—transcend political borders and national political institutions. Inventing radical uses for digital technologies, members of TANs disseminate information rapidly, frame issues, mobilize support, and accomplish effective lobbying of intergovernmental organizations. TANs are meeting individual and collective information needs that have been poorly understood by the mainstream media. Andrew Kohut, director of the Pew Research Center for the People and the Press, says that "[p]ublic interest [in international news] is high and unmet [by the mainstream media], but that interest is less in politics and the stuff of governments than it is in, say, global warming, or hoof-and-mouth disease or the status of women."[31]

Keck and Sikkink—who have written one of the most insightful studies of the increasing role of advocacy networks in international politics—suggest that these new transnational formations are "communicative structures" because "at the core of network activity is the production, exchange, and strategic use of information."[32] In the 1990s, running an online network became considerably less costly, and thus these mushroomed. One can appreciate the rapid growth of online networks by looking at organizations like AlterNet, which was started in 1998 and six years later has "an average of over 1.7 million visitors each month." In addition, its daily newsletter reaches over 90,000 people and its blogs have "more than 80,000 readers per week."[33] AlterNet defines itself as both a "news magazine and online community that creates original journalism and amplifies the best of dozens of other independent media sources." The fact that AlterNet sees itself as both a medium and a community is quite revealing of the essential role that communication plays in TANs.

Members of TANs are not motivated by financial profit, and their work is structured by Benkler's commons mode of production. The radical (or citizen) journalism that they create seeks to bring new issues to the global public sphere, to frame old issues in inventive ways, and ultimately to persuade global audiences and pressure governments and intergovernmental organizations. Keck and Sikkink say that TANs transform the "information and value context within which states make policy" by engaging in four types of tactics: information politics, symbolic politics, leverage politics, and accountability politics.[34] The first two tactics are particularly important for understanding the ongoing change in the production and distribution of "foreign news." Information politics includes informal exchanges through e-mail, fax, and telephone, as well as small media outlets, especially Web sites, pamphlets, and newsletters. Enabled by the Internet to reach millions across borders, TANs have become significant players in the production and distribution of transnational news. Oftentimes, acting as amateur investigative reporters, activists gather information and write stories themselves, but as Keck and Sikkink note, "to be credible, the information produced by networks must be reliable and well documented."[35]

A second crucial component of TANs' information politics is the framing of issues in ways that show the injustice of a given situation and put forward compelling solutions. Stories are told in plain rather than technical language, and the wrongdoers are clearly identified. A telling example cited by Keck and Sikkink is the movement against female circumcision, in which the naming of the practice became a major point of contention in the international debate. Women's movement advocates rejected technical terms like *clitoridectomy* and *infibulation* and framed the issue in terms of violence against women by renaming the practice "female genital mutilation."[36] Scholars like Todd Gitlin have pointed out that news produced by activists is circulated mainly within their networks and that such news does not directly enrich the public sphere, but rather the networks create their own public "sphericules."[37] However, advocacy network news frequently reaches broader audiences. The networks of major groups such as Amnesty International, Earth Rights International, the International Feminist Network, and the International Confederation of Free Trade Unions publish timely reports and constantly generate news that often finds its way into mainstream media. A critical component of their information politics is to attract press attention. "Sympathetic journalists may become part of the network," say Keck and Sikkink, "but more often network activists cultivate a reputation for credibility with the press, and package their information in a timely and dramatic way to draw media attention."[38] The success of the campaign to ban land mines is an excellent example of this; advocates were able to create global awareness by enlisting celebrities as spokespersons and capturing the mainstream media's attention. Moreover, the second tactic of TANs, symbolic politics, supports efforts to attract press coverage. Sometimes activists attach a symbolic meaning to major international events, constructing an interpretation of the event that has significance for publics outside the network. For example, environmentalists transformed the Brazilian rainforest burning in 1988 into a symbol of global warming. Deploying spectacular footage, they persuaded many ordinary people in the United States to support activist views regarding the larger political issues surrounding global warming.[39] A more recent example of symbolic politics is the antiwar movement's astonishing deployment of the Abu Ghraib prison photos (a Google search yielded about 6,470,000 results for "Abu Ghraib" and about 142,000 for "Abu Ghraib photos"). As sociologist Paul Starr says, "When other aspects of the Iraq War have long been forgotten, the images of American soldiers torturing Iraqis in Abu Ghraib prison will still be remembered."[40]

While most TANs, as politically-engaged transnational communities, seek the core values of liberal democracies, there are other, very different transnational networks that have become major actors in "foreign" affairs reporting. Most notably, Al Qaeda can be seen as a transnational community. "For the United States Department of Defense," says Vertovec, "transnationalism means terrorism, insurgents, opposing factions in civil wars conducting operations outside of their country of origin, and members of criminal groups."[41] Looking at "foreign news" from the perspective of transnational networks, what has changed is the nature of foreign news itself.

The emergence of the "new foreign news" that Powers describes has coincided with the largest migratory movement in history. Ethnic groups like people of Chinese, Turkish, or Punjabi origin or descent living in North America and Western Europe are yet another type of transnational community that produces, distributes, and uses huge amounts of transnational news. To give a sense of the nuances of these transnational news practices, in the remainder of this chapter we talk about U.S. Latinos because they are the largest minority group in the United States and they engage in deeply transnational media practices.

## THE INTENSE TRANSNATIONALISM OF LATINOS

Here we focus on two dimensions of news in transnational spaces. One is regarding Latinos themselves as transnationals, what that means for their interest in news, and what might be considered "foreign" news to them. The second dimension looks at transnationalism as inherent in the Latino media's news itself, as it is produced for an audience that includes many transnational viewers. Latinos account for approximately 13 percent of the total population of the United States. Perhaps most important for commercial media systems is the size of the Hispanic market, which is estimated to have a buying power of $715 billion and is expected to double in the next decade.[42] Most Latinos identify themselves by their country of origin or ancestry, although increasingly many also accept the panethnic term *Latino.* There is enormous diversity within this group in terms of language (not all Latinos speak Spanish), race, class, citizenship/residency status, religion, and nationality. The group includes undocumented workers as well as elite professionals; U.S.-born citizens, labor migrants, and political refugees; Catholics, Protestants, Jews, and non–religiously affiliated people. Although it is frequently claimed that Spanish, Catholicism, and some form of immigration experience unite them, these characteristics certainly do not describe all Latinos.

Latino transnationalism is a sound model for the way many ordinary people relate to news about events that crisscross political borders. The homeland is always subjectively close to the diaspora, but in the case of Latinos, geographic proximity and advances in transportation constantly reinforce attachment to "home." Such profound attachment drives the huge remittances and investments that, in countries like El Salvador, became the largest source of foreign exchange in the 1990s. Immigration scholars insist that there is a new pattern of immigration: the transnational migratory circuit. "This pattern of immigration," write scholars Marcelo M. Suárez-Orozco and Mariela M. Páez, "is typified by intensive back and forth movement, not only of people but also of goods and information."[43] For example, sociologist Robert C. Smith details the myriad ways that two generations of Mexicans from Ticuani, Puebla, living in New York City live transnational lives. Besides economic remittances that sustain the town of Ticuani, Smith says that "the most striking aspect of Ticuani transnational life is that the Ticuani municipal authorities in Mexico and their immigrants' counterparts in New York have worked out shared administrative and political arrangements to carry out public works and other projects back in the home town, syncretically adapting older institutions to the migrant context."[44]

## EFFECT OF MAINSTREAM FOREIGN COVERAGE ON LATINOS AND OTHER GROUPS OF NON-EUROPEAN ORIGIN

Not all Latinos immigrated to the United States. Many are the U.S.-born children of immigrants or trace their foreign ancestry several generations back. Puerto Ricans became U.S. citizens when the United States annexed the island in 1898, and many Mexican Americans are descendants of Mexican settlers whose lands became part of the U.S. territory in 1848. Yet like Asian Pacific Americans, Arab Americans, and other groups of non-European origin, often all Latinos are perceived as foreigners and as having divided national loyalties. There are many reasons for this public perception, as Latinos have a very mixed history of integration into the U.S. mainstream society, but the media may have contributed to it. Recent studies of Latino representation in network news continue to show that Latinos appear less than 2 percent of the time.[45] Near invisibility in the mainstream media, when added to the few existing images that often portray Latinos as foreigners, may have fostered the perception that they do not belong to this country.

Being seen as foreigners has implications for the way Latinos relate to foreign news as well as for the way Latino journalists present foreign news. The media coverage of their country, or even region, of origin or descent can have serious consequences for the presumed foreign individuals and communities. A notorious example is the case of Vincent Chin, a Chinese American who was beaten to death in 1982 in Detroit, in the midst of the largely negative media coverage of Japan that followed Japanese automotive companies' success in the U.S. market. He was brutally murdered by two white autoworkers who, mistaking him as Japanese, blamed him for the loss of their jobs. There are more recent examples. In the aftermath of 9/11, Arab Americans became keenly concerned about the coverage of the Middle East. A press release posted on the American-Arab Anti Discrimination Committee's Web site shows their concern "by the alarming hostility expressed by media commentators towards the Palestinian people in the wake of the death of Palestinian Leader Yasser Arafat.[46]

Stereotypical portrayals and scarcity of news about their U.S. communities and their "home" countries in the mainstream media are two of the reasons minority groups seek information in alternative news sources. The growing popularity of ethnic media is in part due to their international news coverage. The first-ever survey of ethnic media in California found that "news television programs are most popular with Californians that watch ethnic television . . . and international news is the most important to those who read an ethnic newspaper."[47] Ethnic audiences often develop a "dual vision" regarding public affairs because they attend to media with different perspectives. For example, a survey by the Pew Hispanic Center notes that "nearly half of the adult Hispanic population crisscrosses between . . . [English- and Spanish-language news], getting some of its news in both languages."[48]

## USE OF NEW INFORMATION AND COMMUNICATION TECHNOLOGIES AMONG LATINOS

A 2004 survey by the Pew Hispanic Center confirmed what previous research has found: most Latinos do not get their news through the Internet. "Only 29% of the adult Hispanic population gets news on the Web," states the survey report.[49] A focus-group study of Latino adults also found that Internet usage was low among them. However, the study also notes that much of these Latinos' knowledge of computers and the Internet came from their children, who were frequently mentioned as being Internet users. So it may be only a matter of time before Latinos become heavy users of online communication. Indeed they use other new communication technologies profusely, as telephone companies know well. Paul Leonardi, the author of the focus-group study, says that Latinos expressed overwhelming enthusiasm for, and familiarity with, cell phones, which they claimed help them stay in touch with family more closely.[50]

Another technology that is central for most Latinos as information seekers is satellite television. Some turn to the Spanish-language media because they are not proficient in English. Yet many Latinos attend to these media because they are looking for a perspective that seems more fair and accurate to them than the mainstream media's usual coverage of corruption, violence, drugs, and natural disasters. Latinos have developed a vibrant media of their own, which often covers events happening in Latin American and Caribbean countries more as domestic than as foreign affairs. The single most important outlet of this news is Univisión, the largest U.S. Spanish-language television network.

## UNIVISIÓN: TRANSNATIONAL FROM THE BEGINNING

Univisión is a transnational company serving a transnational audience, and in the process turning a traditional U.S. notion of "foreign news" on its head. It has been transnational since its inception—long before the words "globalization" or "transnational" became part of our vocabulary.

The origins of Univisión are the Spanish International Network (SIN), which was launched in 1961 by Televisa, the Mexican media conglomerate. Through SIN, Televisa started to dominate a segment of the U.S. market. As América Rodríguez, a scholar who did an extensive study of Univisión's newscasts, states, "This unusual situation—that of a developing country annually exporting thousands of hours of television programs to a 'First World' country—provides an early example of the transnational nature of an emerging global media economy."[51]

SIN began with two stations, one in San Antonio and one in Los Angeles, but soon expanded. The programming came exclusively from Mexico in the form of *telenovelas* (soap operas) and Mexican films, while the advertising base came from the United States. For Mexicans living in the United States this programming provided a nostalgic view of the homeland and kept them connected with Mexico in their hearts if not always in their bodies. By the 1970s, SIN was also broadcasting Televisa's *24 Horas,* the nightly national newscast, from Mexico.

SIN was well positioned to expand into New York and Miami when large numbers of Puerto Ricans and Cubans began establishing themselves in these cities, respectively, in the 1970s. There was literally no other Spanish-language television available because SIN had continuously bought out potential rivals. SIN was able to rapidly expand by being an early adopter of new television technologies such as the UHF band, cable, microwave, and satellite interconnections. For example, its growth into cable began in 1972, eight years before CNN was launched. The use of microwave technology to interconnect stations in the west allowed the sale of regional advertising. And "in 1976, SIN became the first U.S. broadcaster to distribute its signal by satellite."[52]

After several changes in ownership and one name change, Univisión is currently 50 percent foreign owned—Televisa still owns 25 percent, and the remaining 25 percent is owned by the Venezuelan conglomerate Venevisión. Univisión is a publicly traded company that owns and operates eighteen full-power and nine low-power stations in the continental United States plus two stations in Puerto Rico. In addition, the network is distributed through 61 broadcast affiliates and 1,890 cable affiliates in the country.[53]

Today, Univisión consistently ranks as the fifth-most-watched network in the United States. It has been capturing market from the major U.S. broadcast networks. In May of 2005, Univisión's Dallas station, KUVN, "was consistently the #1 station among Adults 18–34, regardless of language, during prime time, Sign-on-Sign-Off, Early Fringe, Daytime and also had the most watched local newscasts at 5 pm and 10 pm."[54] The same thing happened in Miami in July.[55] And in a major coup, Univisión's local and world news *(Noticiero Univisión)* broadcasts beat the other networks in New York City four times in two weeks in August 2005. This was considered even more remarkable given that Census data puts New York's television market at only 19 percent Hispanic.[56] Despite the fact that Univisión has been the target of strong criticism by academics and Latino advocacy groups, many Latinos support its existence and efforts because it provides a powerful symbol of their visibility in U.S. society.[57] The Pew survey found that 61 percent of those Latinos who said they get their news in English ranked the Spanish-language media as "very important to the economic and political development of the Hispanic population."[58]

## *NOTICIERO UNIVISIÓN*

"Network television coverage of news from Latin America is the stronger draw for any Spanish-language media," writes Roberto Suro, director of the Pew Hispanic Center.[59] SIN created its nightly newscast, *Noticiero SIN,* in 1981, later renaming it *Noticiero Univisión.* It would become a winner for the network. Even Latinos who are critical of some of the entertainment programming of Univisión often watch and appreciate its flagship newscast.[60]

Even though it was created by a Mexican conglomerate, *Noticiero* developed a unique style, which is quite different from Televisa's Mexican newscasts. Rodríguez says that the professional tenet of objectivity made Univisión's journalists U.S.—rather than Latin American—journalists. This objectivity also gave *Noticiero* credibility both among Latinos and among other U.S. networks.[61] Rodríguez also found that by launching a national newscast Univisión aimed to build up a national, panethnic U.S. Latino audience, as opposed to the existing regional audiences made up of single groups based on national origin (i.e., Cubans, Mexicans, and Puerto Ricans). This panethnic goal actively guides production decisions. For example, there is an assumption that Mexican Americans in California will be interested in political events in Cuba just out of concern for fellow Latinos, not because they personally have a vested interest in the politics of Cuba. A related assumption is that Latinos will be more interested in news about other Latinos than in news about other groups in the United States.

While this is a guiding tenet of *Noticiero,* the popular male anchor Jorge Ramos tells of an incident that reveals how Latin American and Caribbean foreign policies are played out in U.S. Latino media. After being moved from Los Angeles to Miami early in his career with Univisión, he came under attack by Cuban exile radio commentators who were indignant that a Mexican would be hosting a show in Miami, given the Mexican government's close and continued friendship with Castro's regime in Cuba.[62]

Univisión provides a Latino cultural lens to the news of the day, and it includes more news than the mainstream networks about both U.S. Latinos and their countries of origin or ancestry. To take a recent example, in the extensive coverage of the aftermath of Hurricane Katrina that hit New Orleans, along with ABC's *World News Tonight, Noticiero* broadcast a story about people not wanting to evacuate the city because they did not want to leave pets behind. Other stories that the networks shared included an address given by President George W. Bush and actions of the Federal Emergency Management Agency (FEMA). However, immediately after the first commercial break *Noticiero* reported on how the Mexican government was sending its military to assist with the relief effort by providing engineers and mobile kitchen units that could prepare up to seven thousand meals. *Noticiero* also reported on offers of assistance from Puerto Rico and even Venezuela. *World News Tonight* did not carry any reference to these news items from Latin America.

This content resonates with Rodríguez's findings a decade ago. In her comparative content analysis of the two newscasts, she found that while 79 percent of *World News Tonight* was national news, *Noticiero Univisión* dedicated slightly more time to news from Latin America (45 percent) than to U.S. national news (43 percent). "The resultant worldview," says Rodríguez, "is somewhat bifurcated: Univisión's journalists and their imagined audience are of two worlds, not fully removed from one yet not fully a member of either."[63]

The extensive coverage of Latin America and the Caribbean has also to do with the fact that Univisión is fighting to capture the "larger Hispanic market." At present, the network is widely available on cable systems in Latin America. While most Latin Americans cannot afford cable, some watch Univisión news indirectly because in some countries "local television stations often simply tape stories from Univisión's or Telemundo's [the second-largest Spanish-language network in the United States] nightly newscasts for their own use, which gives these American networks a degree of credibility and visibility unusual in the region."[64]

*Noticiero Univisión* fits easily into the Latin American and Caribbean context because, as Rodríguez points out, the newscast is infused with "panamericanism." In the transnational configuration of audiences, *Noticiero Univisión*'s coverage of Latin America and the Caribbean is based on a transnational panamerican map rather than on a map of the United States and outside countries. The audience spans a north-south axis through the United States and down to South America. Foreign news then becomes news about Asia or Europe, not news about Peru or Argentina, and the transnational lives of the audience as well as the transnational production and ownership of the network are validated.

## SUMMARY

Most mainstream news organizations, or at least their parent corporations, seem content to rely on either wire services or parachute journalism to cover events outside the United States. Despite a slight revived interest after September 11, foreign news in the old tradition of foreign correspondence continues to disappear.[65] And this is certainly not because there is no interest on the part of the public, as the tremendous growth of advocacy networks' online media demonstrate. What we are suggesting is that new patterns of transnational living and communicating may require news organizations to make a fundamental shift in how they think about "foreign" news. This suggestion goes a step beyond simply emphasizing local links in global stories. That model is certainly an improvement, but it neglects many of those fascinating stories of a transnational character occurring in transversal spaces. Those stories are worth searching for and providing to readers/viewers.

While there is strong and growing interest in diasporic groups as a market and a great deal of recent scholarly work on transnationalism, these topics seem to be outside the realm of discussion of mainstream journalists and journalism studies. It seems to us that there might be some areas of convergence that are being overlooked in the discussion of foreign news. "Globalization has upset the familiar dichotomy between 'foreign' and 'domestic' affairs," says political scientist James Anderson,[66] arguing that globalization has put transnational democracy on the agenda of political science. Paraphrasing his argument, we suggest that the category "transnational news" needs to be added to the journalistic agenda.

NOTES

1. Steven Vertovec, "Conceiving and Researching Transnationalism," *Ethnic and Racial Studies* 22, no. 2 (1999): 447.

2. Karim H. Karim, From Ethnic Media to Global Media: Transnational Communication Networks Among Diasporic Communities (Hull, Quebec: Department of Canadian Heritage, 1998), 14.

3. Ibid., 11.

4. UN Department of Economic and Social Affairs, "World Economic and Social Survey: International Migration," http://www.un.org/esa/policy/wess/wess2004files/part2web/preface.pdf.

5. The terms *Latino* and *Hispanic* refer to U.S. residents who trace their origin or descent to Latin America and the Caribbean. Residents of these two regions are called "Latin Americans" and "Caribbeans."

6. Marcelo M. Suárez-Orozco and Mariela M. Páez, "Introduction: The Research Agenda," in *Latinos Remaking America,* ed. Marcelo M. Suárez-Orozco and Mariela M. Páez (Berkeley: University of California Press, 2002), 11.

7. Don Flournoy, "Coverage, Competition, and Credibility: The CNN International Standard," in *Foreign News: Perspectives on the Information Age,* ed. Tony Silvia (Ames: Iowa State University Press, 2001), 37.

8. James F. Hoge, "Foreign News: Who Gives a Damn?" *Columbia Journalism Review* 36, no. 4 (November–December 1997): 48–.

9. Dwight L. Morris and Associates, *America and the World: The Impact of Sept. 11 on U.S. Coverage of International Affairs* (Washington, DC: Pew International Journalism Program, 2002), on line at http://www.pewtrust.com/pdf/vf_pew_intl_fellows_911.pdf.

10. Stephen B. Crofts Wiley, "Rethinking Nationality in the Context of Globalization," *Communication Theory* 14, no. 1 (2004): 78.

11. Glen Guzzo, "Thinking Big: Covering Major International Stories Can Pay Significant Dividends for Regional Newspapers," *American Journalism Review* 26, no. 3 (June 2004): 20.

12. William Powers, "Hello, World," *National Journal* 33, no. 26 (2001): 2082.

13. American Society of Newspaper Editors, *Bringing the World Home: Showing Readers Their Global Connections,"* American Society of Newspaper Editors, July 23, 1999, updated January 10, 2000, http://www.asne.org/index.cfm?ID=2569.

14. John B. Alterman, "Transnational Media and Regionalism," *Transnational Broadcasting Studies* 1 (Fall 1998), http://www.tbsjournal.com/Archives/Fall98/Articles1/JA1/ja1.html.

15. David Shaw, "Foreign News Shrinks in Era of Globalization," *Los Angeles Times,* September 27, 2001.

16. Leonard Downie, quoted in Michael Parks, "Foreign News: What's Next?" *Columbia Journalism Review* 40, no. 5 (January–February 2002): 52.

17. Michael Grade, prologue, *Future of the BBC,* BBC Web site, http://www.bbc.co.uk/thefuture/bpv/prologue.shtml, 2.

18. Richard Tait, "The Future of International News on Television," *Historical Journal of Film, Radio and Television* 20, no. 1 (2000): 51.

19. Lucinda Fleeson, "Bureau of Missing Bureaus," *American Journalism Review* 25, no. 7 (2003): 36.

20. Ibid.

21. Joe Strupp, "Nouveau Foreign Correspondents' Tastes Range from Wine to War," World Press Institute Web site, http://www.worldpressinstitute.org/internat.htm.

22. Foreign Press Association of New York, "Writing about Everything That Is Worth Some News," Foreign Press Association of New York Web site, http://www.foreignpressnewyork.com/pages/members_reporting.htm, accessed September 11, 2005.

23. James Anderson, "Questions of Democracy, Territoriality and Globalization," in *Transnational Democracy: Political Spaces and Border Crossings,* ed. James Anderson (London: Routledge, 2002), 11.

24. David D. Perlmutter, "Bird Flu Blogging: Truth to Power," PolicyByBlog Weblog, January 9, 2006, http://policybyblog.squarespace.com/journal/2006/1/7/bird-flu-blogging-truth-to-power.html.

25. Vladimir van Wilgenburg, "Turkish State Not Helping Kurds Dying from Flue," Weblog entry, http://vladimirkurdistan.blogspot.com/2006/01/turkish-state-not-helping-kurds-dying.html.

26. Vertovec, 453–54.

27. John Maxwell Hamilton and Eric Jenner, "Foreign Correspondence: Evolution, Not Extinction," *Nieman Reports* 58, no. 3 (2004): 99.

28. Yochai Benkler, *The Wealth of Networks: How Social Production Transforms Markets and Freedom* (New Haven: Yale University Press, 2006), 1.

29. Ibid., 9.

30. World Resources Institute, "Civil Society: International Non-Governmental Organizations with Membership" (table), World Resources Institute Earth Trends online database, 2005, http://earthtrends.wri.org/text/environmental-governance/variable-575.html. Accessed January 2007.

31. Andrew Kohut, quoted in Parks, 52.

32. Margaret E. Keck and Kathryn Sikkink, *Activists beyond Borders: Advocacy Networks in International Politics* (Ithaca, NY: Cornell University Press, 1998), x.

33. AlterNet, http://www.alternet.org/about.

34. Keck and Sikkink, 16.

35. Ibid., 19.

36. Ibid., 67.

37. Todd Gitlin, "Public Sphere or Public Sphericules?" in *Media, Ritual and Identity,* ed. Tamar Liebes and James Curran (London: Routledge, 1998).

38. Keck and Sikkink, 22.

39. Ibid., 22–23.

40. Paul Starr, "The Meaning of Abu Ghraib," *American Prospect,* June 2004, http://www.princeton.edu/~starr/articles/articles04/Starr-MeaningAbuGhraib-6-04.htm.

41. Vertovec, 450.

42. Stephanie Pillersdorf and Brooke Morganstern, "Univision Announces 2005 Second Quarter Results," Univision Web site, August 4, 2005, http://www.univision.net/corp/en/ir/2q05.pdf.

43. Marcelo M. Suárez-Orozco and Mariela M. Páez, "Introduction: The Research Agenda," in *Latinos Remaking America,* ed. Marcelo M. Suárez-Orozco and Mariela M. Páez (Berkeley: University of California Press, 2002), 11.

44. Robert C. Smith, "Comparing Local-Level Swedish and Mexican Transnational Life: An Essay in Historical Retrieval," in *New Transnational Social Spaces: International Migration and Transnational Companies in the Early Twenty-first Century,* ed. Ludger Pries (New York: Routledge, 2001), 43.

45. Federico Subervi, *Network Brownout 2004: The Portrayal of Latinos and Latino Issues in Network Television News* (Austin and Washington, DC: National Association of Hispanic Journalists, 2004), 5.

46. American-Arab Anti Discrimination Committee, "Protest Biased Media Coverage of Palestine and Palestinians," American-Arab Anti Discrimination Committee Web site, November 13, 2004, http://adc.org/index.php?id=2383.

47. New California Media, "First-Ever Quantitative Study on the Reach, Impact and Potential of Ethnic Media," American Media Web site, April 2, 2002, http://news.ncmonline.com/news/view_article.html?article_id=796.

48. Roberto Suro, "Changing Channels and Crisscrosing Cultures: A Survey of Latinos and the News Media," Pew Hispanic Center, http://pewhispanic.org/files/reports/27.pdf, 2.

49. Ibid.

50. Paul M. Leonardi, "Problematizing 'New Media': Culturally Based Perceptions of Cell Phones, Computers, and the Internet Among United States Latinos," *Critical Studies in Media Communication* 20, no. 2 (June 2003): 160–79.

51. América Rodríguez, "Creating an Audience and Remapping a Nation: A Brief History of U.S. Spanish-Language Broadcasting, 1930–1980," *Quarterly Review of Film and Video* 16, no. 3–4 (1999): 365.

52. Ibid., 366.

53. "Univision Media Properties: Univision and Telefutura Television Groups," Univisión Web site, http://www.unvision.net/corp/en/utg.jsp.

54. Colleen Carnahan, "KUVN-TV 23 Reigns as Top Rated Station in Dallas for Adults 18–34," Univisión Web site, http://www.univision.net/corp/en/pr/Dallas_20062005-1_print.html, 1.

55. Brooke Morganstein, Cassandra Bujarski, and Shannon Provost, "Univision Shakes up ABC, CBS, NBC and FOX in Historic 2004–2005 Season," Univisión Web site, http://www.univision.net/corp/en/pr/Los_Angeles_02052005-2.html.

56. Frankie Miranda, "Univision News #1 in Gotham, 4th Time in Two Weeks," Univisión Web site, http://www.univision.net/corp/en/pr/New_York_04082005-2_print.html.

57. Viviana Rojas, "The Gender of Latinidad: Latinas Speak about Hispanic Television," *Communication Review* 7, no. 2 (April–June 2004): 125–53.

58. Suro, 2.

59. Ibid., 14.

60. Rojas, 132.

61. América Rodríguez, "Objectivity and Ethnicity in the Production of the Noticiero Univision," *Critical Studies in Mass Communication* 13 (1996): 59–81.

62. Jorge Ramos, *No Borders: A Journalist's Search for Home,* trans. Patricia J. Duncan (New York: Rayo, 2002).

63. Rodríguez, "Objectivity," 68.

64. Karim, 14.

65. Morris and Associates.

66. Anderson, 10.

# 3

# THE NOKIA EFFECT

## *The Reemergence of Amateur Journalism and What It Means for International Affairs*

STEVEN LIVINGSTON

### INTRODUCTION

Economic and technological forces have placed tremendous pressure on traditional journalistic practices and norms. On the economic front, the emphasis placed on profit by the corporate news media starves the pursuit of serious international news while it encourages dramatic but otherwise trivial content. As a result, in the last decade overseas bureaus have been shuttered, foreign correspondents sacked, and airtime and print space devoted to international news slashed.[1] As veteran foreign correspondent Tom Fenton wrote in his thoughtful critique of American international affairs news coverage, the "mega-corporations that have taken over the major American television news companies squeezed the life out of foreign news reporting."[2]

Just as disturbing to many news professionals is the realization that the economics of contemporary corporate news constitute only a part of the larger set of challenges confronting American journalism. This chapter considers how new information technology affects contemporary journalism. It also considers questions about how event-driven news—the sort of news favored by new technology—undermines the public's ability to hold policy makers accountable while increasing the likelihood of supporting a more bellicose foreign policy.

### NEW TECHNOLOGY AND NEWS GATHERING

Several new information technologies are pulling viewers and readers away from traditional news media. As a recent report by the Carnegie Corporation noted, "Through Internet portal sites, handheld devices, blogs and instant messaging, we are accessing and processing information in ways that challenge the historic function of the news business and raise fundamental questions about the future of the news field."[3] Those with concerns about the future of journalism would find additional discomfort in the declining numbers of newspaper readers in the United States, the creaky status of the network evening news programs (including the apparent demise of the star anchor system), and recent public opinion survey results revealing that Americans trust the news media less than every other major institution in American life. With these trends and public attitudes as backdrop, it is perhaps not surprising that American news consumers are turning to alternatives offered by new technologies.[4]

Ironically, many of the same technologies that threaten traditional news consumption habits dramatically expand the ability to gather and distribute socially relevant information. Often, this enhanced information gathering capability is used by traditional news organizations when reporting news. Examples of this include personal camcorder images used by television news to report natural disasters, plane crashes, or other events caught on camera. On other occasions, traditional media are bypassed altogether, such as when text messaging was used by Chinese citizens to spread news of the SARS epidemic at a time when official state-controlled media ignored it as a matter of official editorial policy.

What sort of coverage is produced by the contradictory pull of shrinking news budgets on the one hand and the expanded technological capacity to cover events globally on the other? Fenton and other critics of American broadcast television news are probably correct when they argue that corporate news executives' unyielding drive to squeeze profit from less substantive news does not auger well for foreign news coverage by broadcast networks. ABC (Disney), CBS (Viacom), and NBC (General Electric) seem unwilling to make the financial commitments required to recreate and maintain first-rate foreign affairs news gathering operations. For the many Americans who still rely on the broadcast networks for their news—about 35 million each evening—this is bad news.

At the same time, the new technologies used to gather news, in conjunction with the Internet and a growing number of regional satellite television channels, mean that more news is now available to more people around the globe than ever before. For those Americans with satellite or cable television or high-bandwidth Internet connectivity, there has never been a richer menu of news options. Typically, when analysts describe the new riches in news options they focus on the Internet, and for good reason. When surveys ask people where they turn for news, a growing number say the Internet.[5] This is especially true of younger people. Countless books, articles, and reports have intoned this basic proposition. But this almost singular focus on the social and political significance of the Internet discourages closer examination of other important technologies. Rather than focusing yet again on the Internet, this chapter considers the implications of other digital information gathering and distribution devices.

Two general categories of technology come to mind: First are the small, highly mobile, multifunctional handheld devices that are more common and more capable than ever before. Secondly, there are a growing number of satellite television channels serving regional, cultural, and linguistic groupings. Examples include Al Jazeera, Al Arabiya, and Al-Manar in the Middle East; Zee TV and several other satellite channels in South Asia; and Telemundo, Univisión, and Telesur in Latin America.[6] Although regional in orientation (usually owing to language), their impact—particularly that of their pictures—is often global in reach.

Together, the proliferation of information collection devices, access to the Internet, and satellite television channels have changed the nature of international news gathering and consumption. Here is an outline of these changes and their consequences:

1. *Ubiquitous sensors:* If professional journalists are not present to capture an event, someone with a camera, cellular telephone, or digital camera often is. Live event reporting with pictures has become the gold standard of modern reporting. Devices of one sort or another create an environment rich in sensors, a condition I have elsewhere called systemic transparency.[7]
2. *Changing nature of journalism:* Pictures from all sources, often amateur—whether they be of a tsunami crashing ashore, a plane tumbling to earth, or the beheading of a man—are fed to the growing number of news organizations, satellite television stations, and Internet Web sites. This development has changed and—in the view

of some—undermined the standard norms and practices of journalism. A local news mantra, "If it bleeds, it leads," can now be applied to international news. But is a steady stream of uncontextualized pictures and accounts of violence properly called journalism?

3. *Global information network:* Formal and informal relationships among these organizations (such as CNN's relationship with Al Jazeera) mean that the pictures captured by amateurs and used by a news organization, or simply posted on an Internet Web site, become grist for a global media mill of shared images and stories. Although local values and norms are often applied to the use of these images, the same images are, nevertheless, distributed around the globe.
4. *Episodic news:* News defined by the availability of pictures and drama usually lacks thematic context and political or historical perspective. Indeed, many would not consider it news properly understood (see number two above), though it is presented as such on the evening television newscast. Of our theoretical considerations, this is most significant. News driven by events captured by a growing array of devices and fed into the global media undermines the public's ability to think critically about events. Context is lost, pushed aside in an endless series of pictures from obscure places. According to Shanto Iyengar, this sort of news short-circuits the public's ability to assess responsibly the conditions created by policy decisions. Even more significantly, he argues that episodic news encourages acceptance of foreign policy solutions to problems misunderstood as the consequence of "evildoers," overly simplistic individualized problem origins.[8]
5. *Response acumen:* Politically, such an information environment demands greater "response acumen" by policy makers. Put another way, rather than setting an issue agenda, as has typically been the case in matters of international affairs, policy makers must now learn to adapt to and even leverage issues that pop up unexpectedly in the news. When event-driven news becomes the norm, response acumen is essential. Former House speaker Newt Gingrich implicitly endorsed this view when assessing President George W. Bush's handling of the aftermath of Hurricane Katrina. Despite the heavy

criticism concerning the inept response to the crisis, Bush could, according to Gingrich, recover "if [he was] seen to improve, re-energize his plans. Otherwise, it swamps the rest of his agenda."[9] Events are seminal moments. They are either triggering or focusing events to be taken advantage of by entrepreneurial policy makers, or they generate news coverage that invites scrutiny and sharper criticisms.[10]

Space limitations permit us to focus on only two of the five points outlined above: *the changing nature of journalism* and the consequence of *information without context.*

## THE CHANGING NATURE OF JOURNALISM

We will begin our consideration of the changing nature of journalism by reviewing a few of the more significant technological trends driving news gathering practices. Our brief review is intended only to illustrate the point; the emerging information environment is not dependent on any one gadget or device. Instead, it is the product of what Bruce Bimber calls the information ecology.[11] Previous information ecologies were characterized by the necessity of managing the flow of physically stored information by large and hierarchically structured organizations. Information today is abundant and stored and distributed electronically. As a result, organizations are less hierarchical and more networked.[12] Devices of all types create information abundance. What does the new information ecology look like? It looks small, networked, mobile, and expanding.

Computers are moving off the tops of desks and laps and into pockets and purses. The trend is toward lighter, smaller, more efficient energy consumption, and therefore highly mobile devices that bundle multifunctionality into one handy instrument. Telephones, cameras (including video), message texting, synchronized organizers, e-mail, MP-3 players, GPS, and Internet access are now standard components for what were once mere cell phones. For our purposes, their key feature is their mobility and multifunctionality. They both send and receive information of various types to and from dispersed and unfixed locations.

In 2005, for example, Nokia launched a "third-generation smartphone" that integrated two cameras, a flash function, and a screen with a 262,144-color capacity. It had two-way video calling functionality that allowed people to share live camera views or stored video clips during normal voice calls. It also received and sent e-mail. This is just one of the many multifunctional devices found on the market at the time of this writing. It of course goes without saying that within a few years almost any technology described in these pages as "new and cutting edge" will bring a smile to the face of knowing readers. Today's "cutting edge" is tomorrow's eight-track tape deck.

What is of particular interest to us is the growing popularity of cameras bundled with cell phones. In 2004, camera phones outsold digital still cameras by a four-to-one ratio. According to Strategy Analytics, a market research firm, 257 million camera phones were sold worldwide in 2004. The year before, 84 million were sold. In just two years, 341 million camera phones were sold around the globe. In 2005, one trade analyst argued that within a short period, "the typical 1.3 megapixel camera phone will be ubiquitous."[13] Furthermore, sale of these devices is no longer limited to a handful of developed countries. One in four cell phones will be sold in Asia. Even in rural Africa, cellular telephony was becoming more common. Indeed, according to one report, by 2005 the fastest growth rate in cell phone use was in Africa.[14]

Although low-end digital cameras (those below 2 megapixels) are being pushed aside by camera phones, one should not discount conventional digital photography when one considers news and politics. In 2005, manufacturers shipped nearly 67 million units. By 2008, annual sales of digital cameras are expected to reach 100 million units.[15] These are often small, high-quality cameras that are easily carried in pockets or bags.

With the proliferation of camera phones and standard digital cameras, an important new phenomenon emerged: we have entered the age of the pocket paparazzi. This phenomenon was illustrated in 2005 at the funeral of Pope John Paul II. In an attempt to maintain the dignity of the event, Vatican officials issued rules prohibiting the use of "cameras" by the million-plus persons who viewed the body of Pope John Paul II as he lay in state at the Vatican. What Vatican officials didn't realize is that the definition of camera is not what it used to be. Thousands of people took camera phone photographs.[16]

There is a growing probability that if something sensational or visually interesting happens, someone with some sort of recording device will be there to capture it and, subsequently, post it on a Web site or feed it into some other aspect of the global media.

The utility of small mobile communication devices is not limited to taking pictures. Mobile voice and text communication offers eyewitnesses the opportunity to share experiences of an event contemporaneously with a wider audience. Global sales of cell phones of various types and capacities were expected to reach nearly 800 million units in 2005. The annual tally of global cell phone sales in 2009 will, according to 2005 estimates, exceed 1 billion units. Within four years of this writing, industry analysts estimate that 2.6 billion cell phones will be in use worldwide.[17]

But as impressive as these numbers may be, we must keep in mind that my argument does not rest on cellular telephony or digital still photography alone. We could just as well point to the growing capacity of conventional news media to cover events using smaller, more mobile devices, such as videophones. Inmarsat is the backbone of videophone transmissions and other satellite news gathering (SNG) devices. SNG has expanded the reach of live event coverage to more remote and far-flung locations around the globe.[18] As videophone software improves and bandwidth expands, the current production limitations (chiefly the 15-samples-per-second bit rate that produces the ghostly, grainy images associated with videophones) will improve, challenging viewers to discern the difference between images from a videophone linked by an Inmarsat satellite and those from a conventional television transmission using Ku-band satellite uplinks.[19] We could also point to the expanding number and improved capabilities of commercial remote sensing satellites.[20] These devices, orbiting at 423 miles in space, take high-resolution images on the ground as small as two feet across. Anyone with reasonable financial assets and a set of coordinates can acquire these images.[21]

And what if the global cell phone fascination fades as quickly as it has emerged? In this view, it really doesn't matter that both cellular and landline telephony might be replaced by voice over Internet protocol (VoIP), a software protocol that allows users to make and receive telephone calls over the Internet. As Wi-Fi wireless Internet connectivity expands, VoIP telephony will itself become mobile. The point is that, one way or another, the globe is and will remain blanketed by devices that gather and distribute data, whether in the form of voice, video, digital stills, or text.

How does this affect news gathering? News organizations once sent professional correspondents overseas to find the news;[22] now the news is often the product of networked "sensors" carried by nonjournalists, or what we might instead call accidental journalists. Events today, regardless of their remoteness, are often a part of the news because someone is there with a camera or cell phone (and more often cell phone camera). What affect do these devices have on what we understand as news?

The 2005 World Press Photo contest, administered by World Press Photo (WPP), an Amsterdam-based nonprofit organization of professional photojournalists, selected sixty prizewinning photographs from nearly 70,000 entries submitted by more than 4,000 photographers. Both the number of photographs and the number of photographers were records. According to the *New York Times,* the large number of submissions was linked to the "proliferation of cameras in more and more devices, like cell phones, a phenomenon gradually turning everyone into an aspiring shutterbug." WPP, like professional photojournalism more generally, was faced with a problem. Because the WPP contest is open only to professionals, amateur photographs, such as those taken at Abu Ghraib or in the immediate aftermath of the 2004 tsunami, were not given recognition by WPP. As photographer Diego Goldberg, the chairman of the 2005 jury remarked about the Abu Ghraib photographs, "Journalistically they were very important, extremely important, but the organization is called World Press, not 'photography in general.' It's about what is being produced by professionals for the press."[23]

Lack of professional status may bar the amateurs at the gates of awards ceremonies, but it does not prevent them from capturing the most important and politically relevant photographs in the world on their Nokia camera phones. The London subway bombings in July 2005 offer examples. Using his Nokia camera phone, Alexander Chadwick, a subway commuter, took one of the iconographic images of the bombings. News organizations around the world used Chadwick's photograph. The *New York Times* carried it on the front page.

The power of the photograph is found in its immediacy. Its grainy, unfocused quality reinforces its evocative power. One sees a line of passengers making their way down a tunnel illuminated only by the harsh

Chadwick's camera phone photograph of London subway evacuation
nytimes.com and flickr.com

glare of emergency lighting. It is a remote, alien world brought home to us by Chadwick's camera phone.

Chadwick was not alone in recording events that day. Amateurs took nearly all of the images of the bombings and their aftermath. Another amateur photographer took about forty photographs, mostly with his Nokia camera phone, and posted about half of them to a Web site called flickr.com. There, his photographs joined hundreds of others taken by amateurs and used by traditional media when reporting the event.

Of course, the fact that amateurs played a key role in news gathering is not at all unprecedented. Over forty years ago, an amateur filmmaker named Abraham Zapruder captured one of the most gut-wrenching images in American history: a graphic twenty-six-second strip of silent 8-mm film capturing the assassination of President John F. Kennedy. The difference between earlier examples of images captured by amateurs and those of the London subway bombing is found in the sheer number of pictures and the speed of their distribution. Flickr.com, a site owned by Yahoo,

Photograph of a man taking a camera phone photograph
nytimes.com

had more than three hundred bombing photos posted within eight hours of the attacks. The BBC and other news organizations in turn posted some of these images, most notably Chadwick's, on their own Web sites.

Another camera phone picture seen often in the days following the London bombing illustrates my point particularly well. What stands out about this image, besides its chilling reinforcement of the claustrophobic terror many of the passengers must have felt, is the central focal point of the image: it is a picture of a man holding a camera phone aloft to take a picture. It is a picture of a man taking a picture of the scene in the tunnel. There is a solipsistic reinforcement of the presence of self, shareable and affirming, made possible by the ubiquitous camera.

Fundamentally, it isn't the number of images or the speed of their distribution alone that is important. Rather, it is the greater probability that someone will capture images of an event, even in the remotest locations, such as the dark, cavernous reaches of London's subway system, or the inner recesses of an American-run prison in Iraq.

For most people, the iconographic status of the Abu Ghraib photographs mean they no longer require actual viewing to be called to mind. We know them by their informal labels: the hooded figure, the leashed man, the pyramid. For most of the world, these images came to symbolize all that was wrong with the U.S. invasion and occupation of Iraq. The perpetrators of the abuses themselves took the pictures, trophy-photos to be shared among friends. Once again, the ubiquitous digital camera played a central role in telling the story.

In the view of the foreign desk editor of the *Washington Post*, Keith Richburg, the pictures made the Abu Ghraib story. Without the pictures, there wouldn't have been a story. A *Washington Post* metro desk reporter assigned to the Frederick, Maryland, news bureau first obtained the pictures. The National Guard unit deployed to the Abu Ghraib prison came from Frederick, an Appalachian community within the *Post's* regional distribution. Although this is an example of solid enterprise reporting, what is interesting is what happened after the *Post* published the initial photographs. According to Richburg, there was a flood of photographs volunteered by other sources stepping forward with their own pictures. "They were calling in and e-mailing the *Post*, saying, 'You think those pictures are something, you should see these.'"[24] With the growing number of digital devices capable of taking pictures, there is an increased likelihood that images or other accounts of events will surface.

Other politically significant examples of amateur images surfacing to play a central part of a political story include the thousands of amateur photographs and videos of the 9/11 attacks on New York and, more recently, the surfacing of a ten-year-old video taken at Srebrenica in Bosnia.

In July 1995, Serbian military forces rounded up and systematically executed approximately seven thousand Muslim boys and men at what was intended to be a United Nations safe area called Srebrenica. In 2005, following up on its rumored existence, a Serbian human rights worker located a copy of a video made of the execution of six Bosnian Muslim men. She submitted it as evidence at the war crimes trial of former Yugoslav president Slobodan Milosevic at The Hague. In addition to its significance as trial evidence, the video had deep political and psychological significance for Serbians, who had tended to discount claims of the massacre as merely Western anti-Serb propaganda. Played repeatedly on Serbian national television, the video was irrefutable evidence that challenged the "commonly held view among Serbs that the atrocity never took place."[25] Within days of its broadcast on Serbian television, several men pictured in the video were arrested, and increased pressure was put on the architect of the massacre, Gen. Ratko Mladic, to surrender to authorities.

What implications do these and many other possible examples have for the practice of journalism and the conduct of international affairs? Dan Gillmor, founder of Grassroots Media, an organization that promotes what he calls "citizen journalism," said witnesses' photos and online accounts would reshape the role of traditional news media over time.[26] London, the Indian Ocean tsunami, and many other recent events suggest he may be correct. But the question that emerges from Gillmor's assertion is: Would a reshaped role of traditional news media be good for the prospects of democratic self-governance?[27] Secondly, how does new technology affect the nature of journalism? We will next take up this second question and then turn our attention to the question of its effects on democratic self-governance.

## THE FURTHER EVOLUTION OF JOURNALISM

Philip Seib has written an important book about the ways speed and immediacy—made possible by new technologies—have undermined traditional journalism norms and practices.[28] My additional point to Seib's contribution centers on a historical perspective that puts recent changes in context. Modern journalism, though still dominated by large bureaucratic organizations, is developing a parallel structure that, in important respects, looks remarkably like journalism as practiced two hundred years ago. If it is understood properly as journalism at all, it is best described as amateur journalism.

For well over a century, American journalism has been characterized by the activities of a cadre of professional reporters working for large, bureaucratic organizations. News gathering typically involves the assignment of reporters to beats where they interact with government officials working in even larger bureaucracies. In 1973, Leon Sigal found that the overwhelming majority of the news in the *Washington Post* and the *New York Times* had its origins in the various interactions of reporters and officials. Indeed, about 90 percent of the news originated from a news conference, official press release, interview, or some other means of officials communicating with the press. News was what officials said it was. Conversely, less than 2 percent of the news, according to Sigal's findings, concerned spontaneous events.[29] Tom Fenton has made the same point more recently in his devastating critique of U.S. international news coverage. Fenton remarks, "Most of the time, in truth, most of the media take their cues from the government in deciding which foreign stories to cover."[30] While Fenton is no doubt correct, the close bond between official agendas and news agendas is evolving, at least in part because technology has freed news organizations from their dependence on official handouts.

In a 2003 article, Lance Bennett and I reported that the nature and origins of CNN international news coverage changed over the course of the 1990s.[31] As heavy, cumbersome, and operationally demanding analog equipment was replaced by smaller, more mobile, and operationally simplified digital equipment, remote broadcasts of breaking news became commonplace. Indeed, by the late 1990s, as the Inmarsat videophone (operated by a single technician) replaced or supplemented C-band or Ku-band mobile satellite uplinks (operated by up to a half dozen technicians), event-driven news overtook institutional news in frequency.

In 1994, at the cusp of the convergence from analog to digital technology, event-driven news with some sort of official response actually exceeded institutional news, though not by much. Institutional news, consisting of press conferences, interviews of officials, and background briefings, actually became more common over the course of the next few years. By the late 1990s, however, a clear trend toward event-driven news became evident. By 1998–1999, event-driven news, typically live coverage of breaking events from somewhere around the globe, overtook institutionally sourced news. There is no reason to believe that since 2001, with the enormous amount of videophone coverage of the wars in Afghanistan and Iraq, that this trend has abated.

In 2005, Douglas Van Belle and I reported that from the mid-1960s to the mid-1990s, news of disasters from remote locations was more commonly found in the *New York Times* and on American broadcast network news. We attributed our findings to advances in technology.[32] Although predictability of supply, one of the prime motivators for institutionally based news, remains important in determining news content, the ability to meet competing news values—including drama, the desire for compelling visuals, and immediacy—is facilitated by advances in technology.

There are, no doubt, a variety of explanations for the recent trend toward event-driven news. In the 2003 study mentioned above, Bennett and I found a sharp increase in event-driven news in the late 1990s. Much of the event-driven news in 1999 came in the form of coverage of the NATO bombing campaign against Serbian targets. Then new Ku-band uplinks mounted on trucks ringed the perimeter of Kosovo and offered a steady stream of live event news coverage. News of Kosovo was filled with harrowing images of the aftermath of errant bombs, such as when a NATO aircraft bombed a column of refugees in the belief it was a convoy of Serbian military vehicles, and of refugees streaming into Macedonia. The preponderance of coverage of Kosovo originated from the region and centered on events (errant bombs, targets hit and missed, people killed or saved) and not, at least proportionately, from the corridors of power in Washington and Brussels. News was more likely to be about what happened, rather than what an official said happened. To be sure, officials got their turn—*after* the event. Their role was less as agenda setters and more as reactors.[33]

In 1999, CNN first used a videophone to cover an earthquake in Turkey. But it wasn't until the spring of 2001 that the videophone came into its own. In April 2001, CNN used a videophone to broadcast the departure of the crew of a downed American EP-3 surveillance plane from a restricted Chinese airbase. It was thought to be the first ever unauthorized television transmission from Communist-controlled China. After that, the videophone become a standard technology used by news organizations around the world to cover breaking events live.[34] By the fall of 2001, scores of videophones were used by war correspondents in Afghanistan. Unlike those who relied on the military to transport copy from the battlefield during the 1990–91 Persian Gulf War, correspondents in Afghanistan transmitted live (text, jpegs, and video) using the Inmarsat satellite system.[35]

Is event-driven news, particularly news that is driven by the availability of amateur photographs, properly understood as news at all? In the view of some, journalism is defined by an editorial process that is lacking in amateur information gathering, and even in live-broadcast television: News professionals exercise judgment concerning the importance or relevance of events and processes. Journalism is something more than an unfiltered flow of raw information. Seib articulates this position well in this volume when he draws a distinction between seeing and understanding.

That distinction gets to the heart of the challenge facing journalism in the real-time era. Simply *seeing* can be accomplished with extraordinary breadth and precision: technology ensures that. But to help the public *understand* is far more difficult, and it will require the news media to occasionally take their feet off the accelerator as they provide information.[36]

The obsession with immediacy and live pictures gets in the way of actual reporting. Abdullah S. Schleifer, founder of the Adham Center for Television Journalism at the American University in Cairo, notes that reporters have no time to "sit down and write a script that provides a coherent story . . . four intelligent hours after an explosive event, and not just an ignorant five minutes after the event."[37] The raw immediacy of some information delivery systems (including live television) lacks the qualities that define news: a thoughtful, vetted, edited presentation of information that has benefited from the experience of news professionals. In this view, amateurs telling stories or posting pictures on the Internet is not journalism.

Yet it is important to remember that this understanding of journalism is, in the long view, quite recent. The modern era of journalism began in the nineteenth century as a "one-man band undertaking." As historian and sociologist Michael Schudson explains, in the early nineteenth century "one man acted as printer, advertising agent, editor, and reporter."[38] Amateurs assisted him on occasion; friends and relatives described in letters home what they saw in their travels. News of distant places and events was the product of the voluntary actions of amateurs. Today, the new forces of change in journalism are the one-man-band bloggers and other amateurs who, armed with cameras and cell phones, create a steady stream of content for a new type of journalism.

In the 1830s, industrial production techniques enabled newspapers to provide news to a growing urban middle class, giving rise to the first large-scale institutionally based form of journalism: the penny press. Urbanization cut the cost of newspaper distribution while facilitating inducements to literacy. All of these factors were necessary for the establishment of a new understanding of news, one less rooted in the post-Revolutionary War–era emphasis on party-driven political polemics, occasionally salted with commercial information and amateur accounts of distant events. Instead, the penny press emphasized sensational scandals and political intrigue. By the mid-nineteenth century, editors had begun to rely less on informal contacts and more on paid freelance reporters. By the 1880s and 1890s, the steady rise of reporters' income and the growing acceptance (and eventual expectation) of a college degree for such journalists both indicated and helped solidify the new professional status of journalism. It is exactly this trend toward professionalization and bureaucratization of news that has reached its zenith and is now challenged by one-man bands and amateurs. Journalism has come full circle. It remains to be seen whether this new form of an older standard of journalism will replace or simply continue to supplement institutional journalism.

Whether there is a return to amateur journalism or not does not address the question of the political consequences of event-driven, episodic journalism. What are the effects of modern journalism's fascination with dramatic events often brought to light by new technologies?

## INFORMATION WITHOUT CONTEXT

The proliferation of highly mobile digital devices such as cell phones, cameras, camcorders, and the many other means of collecting information—raw events—accentuates what Shanto Iyengar has called "episodic frames." According to Iyengar, framing refers to "subtle alterations in the statement or presentation of judgment and choice problems, and the term 'framing effects' refers to changes in decision outcomes resulting from these alternatives." Iyengar further specifies two framing subtypes: *Thematic framing* places issues in a broader or more general context. History, culture, and thick descriptions of politics constitute the essence of thematic framing. *Episodic framing,* on the other hand, "takes the form of event-oriented reports and depicts public issues in terms of concrete instances." As noted by Iyengar speaking of television news at the beginning of the 1990s, most television news reports focus on "concrete acts and breaking events."[39] Quoting David Altheide, Iyengar underscores the importance of framing to political outcomes: "Television reports that rely on visuals of an event will be more entertaining to an audience, yet provide little useful narrative interpretation to understand the broader issue. As long as more dramatic visuals are associated with the tactics and aftermath of terrorism, these aspects will be stressed over the larger issues of history, goals, and rationale."[40]

In a series of experiments designed to test this supposition, Iyengar found that episodic framing encouraged a tendency to attribute responsibility for political outcomes to individual rather than more general or societal causes. For example, when stories about politically motivated violence were contextualized by information about local politics, history, and social and economic conditions, subjects tended to attribute responsibility for the violence to those conditions. Conversely, when violence was reported as an isolated act or event, subjects tended to attribute responsibility to the individual perpetrators of the violence. The causal attribution is to "evildoers" rather than more complex social, historical, or political explanations.

In content analyzing some eleven hundred American broadcast network news stories about crime and some two thousand about terrorism in the 1980s, Iyengar found that both topics were "almost exclusively in episodic terms."[41] What are the consequences of episodic news? As Iyengar summarized, "The dominant episodic frame in network coverage encouraged viewers to attribute causal responsibility for terrorism to the personal qualities of terrorists and to the inadequacy of sanctions. Episodic framing also made viewers more likely to consider punitive measures rather than social or political reform as the appropriate treatment for terrorism."[42]

Iyengar concludes that his findings suggest that episodic news coverage tends to encourage public support for a more bellicose foreign policy, one that addresses problems with military might intended to physically eliminate evildoers rather than addressing complex political conditions. As the pace of event-driven news increases, as surely it must with the growing capability to cover events live from around the world, viewers' ability to think through more complex arguments behind or at the core of events erodes. In this way, event-driven news and political sophistication may be inversely related. If more gadgets lead to an increased tendency to report event-driven news, prospects for critical involvement from citizens seem dim. Even before the proliferation of smaller, mobile, and nearly ubiquitous devices that enable the capture of events, television news tended to emphasize episodic frames. Recent advances in technology might exacerbate this tendency.

What is needed to counter episodic frames? According to Tom Fenton, what is needed is a foreign correspondent who can read through the miasma and see meaning in the complexities of the situation. But not everyone is so sanguine about the abilities of American foreign correspondents, even dedicated ones like Fenton, to offer context and sophisticated analysis. From the perspective of those in the areas where news takes place, American foreign correspondents are often hopelessly myopic, nationalistic, and too often simply wrong. Some veteran correspondents are just as likely to bring cultural blinders to the task of reporting international news as they are an enlightened perspective. Officials and pressure groups will also encourage the adoption of culturally congruent frames and punish noncongruent frames. So what is the alternative?

Most of the world gets international news that differs significantly from American news or from the American point of view. Sometimes it is true that foreign news is dismissed stateside as mere anti-Americanism. International audiences are regularly exposed to significantly different interpretations of U.S. policies and operations—interpretations that simply don't surface in American mainstream news media.[43]

An interesting trend is developing in the United States: There is a growing appetite for news from broadcast services originating outside the United States. Thirty-seven million Americans claim to view BBC news each week, mostly the BBC World as carried by PBS stations.[44] Additionally, according to 2005 estimates offered by Nielsen Media Research, there were 11.2 million Latino homes with televisions in the United States.[45] Though I am unaware of empirical research that verifies this supposition, it seems reasonable to believe that Spanish-language news, particularly programming by Telesur, an initiative of Venezuelan president Hugo Chavez, does not match in frame or emphasis the programming found on English-language television news.[46] There is, in short, an expansion of choices available to news consumers. Whether choice will include the possibility of news relatively free of event-driven news remains to be seen. It could just as well be the case that Latino regional satellite news channels will offer a steady diet of events, just as the North American news channels do. Continuing research must provide the answers.

In the meantime, how should policy makers and news professionals respond to this new information environment? What advice might we give to American news professionals?

American news organizations must rediscover the fundamental importance of international news gathering. In the process, they must avoid confusing genuine international news with mere coverage of U.S. foreign policy. Too often, what passes for international news is little more than a singular fixation on the most recent focus of U.S. interests. The world is more complex than that. If the news is to alert the American public and policy makers to developing issues and incipient crises, news must decouple itself from its near exclusive focus on the United States abroad. The BBC is a good model in this regard. Reopen shattered bureaus; reinvest in a cadre of foreign correspondents who are trained in the languages, politics, and culture of the places they cover. This is vitally important if we are to avoid the pitfalls of episodic news.

What about the fixation on the immediate? The commercial imperatives of the news business mean the drama and immediacy of breaking news coverage—facilitated by new technologies—will continue to fascinate audiences and excite news executives. This is Philip Seib's point elsewhere in this book. Understanding is too often sacrificed to the adrenaline rush of speed. But speed, in and of itself, is not the problem, as long as there is equal time given to creating understanding. If American corporate media were to reverse the decline in their commitment to true international coverage, something much closer to Iyengar's thematic frames would be found in the news. This will enable viewers and readers to contextualize news and undermine the tendency to lean on the simplistic understandings that emerge out of more episodic frames.

What can policy makers do? Answering this question fully would require at least another complete chapter.[47] In some measure policy makers are already doing what they can. If one visits almost any office in the White House (save the Oval Office) or Pentagon, one sees television monitors tuned to a twenty-four-hour news channel. At times, television news serves as the most reliable source of information, at least for fast-breaking events. Open-source intelligence is made richer by the growing array of "news gathering " resources around the globe, whether they be new satellite news channels or bloggers. Yet as *New York Times* columnist Nicholas D. Kristof recently pointed out, "The intelligence community has historically downplayed Osint [open-source intelligence]. Robert Gates, the former C.I.A. director, once told me ruefully that intelligence is sometimes undervalued if it hasn't been stolen."[48] To rectify this problem, the Central Intelligence Agency recently opened the Open Source Center,[49] which gleans information from Web sites and other non-classified sources of information from around the globe.

But these suggestions are just the tip of the iceberg. The revolution in information technology is ushering in new power relationships, new organizational structures, and new understandings of what constitutes journalism. It is also changing the nature of warfare, trade, and international affairs. My few closing observations do not do justice to the challenges confronting policy makers and traditional journalism. I hope that my observations concerning the nature of the challenge are themselves sufficient in their contribution to understanding.

NOTES

1. Arthur Kent, *Risk and Redemption: Surviving the Network News Wars* (Toronto: Viking, 1996).

2. Tom Fenton, *Bad News: The Decline of Reporting, the Business of News, and the Danger to Us All* (New York: Regan Books, 2005), 11.

3. Merrill Brown, "Abandoning the News," *Carnegie Reporter* 3, no. 2 (Spring 2005): 3.

4. Ariana Eunjung Cha, "Do-It-Yourself Journalism Spreads; Web Sites Let People Take News into Their Own Hands," *Washington Post*, July 17, 2005.

5. The growing importance of Internet-based news was illustrated recently when four bombs killed over fifty people in London. On July 7, 2005, the day of the attacks, the BBC Web site recorded an all-time bandwidth peak of eleven gigabytes at 12:00 midnight GMT. At peak times during the day, the BBC Web site received forty thousand page requests *per second*. In total, it received approximately one billion total hits on the day of the bombing.

6. As these alternative media proliferate, news will reflect the concerns of the audiences for whom it is intended. This may exacerbate the negative world opinion of the United States. In this volume, however, Margaret DeFleur suggests that U.S.-based media are actually most responsible for the negative assessments of America found in so many parts of the world. U.S.-based media portray America as a self-obsessed, violence-prone society with few redeeming qualities. Whether this is the main source of the negative sentiment or merely one of a number of sources, including certainly American foreign policy, is debatable. Yet it seems safe to say that regional satellite stations such as Al Jazeera do not help generate more positive attitudes concerning the United States.

7. Steven Livingston, "Transparency and the News Media," in *Power and Conflict in the Age of Transparency*, ed. Bernard Finel and Kristin Lord (New York: Palgrave, 2000), 257–85.

8. Shanto Iyengar, *Is Anyone Responsible? How Television Frames Political Issues* (Chicago: University of Chicago Press, 1991).

9. Quoted in Brian Knowlton, "Bush Faces Rising Complaints about Handling of Disaster," *New York Times,* September 4, 2005.

10. The concept of a triggering event as opportunity was formulated by Roger W. Cobb and Charles D. Elder, *Participation in American Politics: the Dynamics of Agenda Building* (Boston: Allyn and Bacon, 1972). See also John W. Kingdon, *Agendas, Alternatives, and Public Policies* (1984; New York: HarperCollinsCollege, 1995); Frank R. Baumgartner and Bryan Jones, *Agendas and Instability in American Politics* (Chicago: University of Chicago Press, 1993). The concept of event-driven news as an invitation for greater scrutiny was most clearly articulated by Regina Lawrence, *The Politics of Force: Media and the Construction of Police Brutality* (Berkeley: University of California Press, 2000).

11. Bruce A. Bimber, *Information and American Democracy: Technology in the Evolution of Political Power* (New York: Cambridge University Press, 2003).

12. William A. Owens with Edward Offley, *Lifting the Fog of War* (New York : Farrar, Straus and Giroux, 2000).

13. Mobiledia, "Camera Phones to Steal Low-End Digital Camera Market," Mobiledia Web site, http://www.mobiledia.com/news/34302.html.

14. "A New Front in Phone Fight," *International Herald Tribune,* August 27–28, 2005.

15. Mobiledia.

16. Elisabeth Rosenthal, "The Cellphone as Church Chronicle, Creating Digital Relics," *New York Times,* April 8, 2005.

17. "Cell-Phone Sales to Reach 779 million This Year," MSNBC Web site, July 20, 2005, http//www.msnbc.msn.com/id/8641618/.

18. Steven Livingston and Douglas Van Belle, "The Effects of New Satellite Newsgathering Technology on Newsgathering from Remote Locations," *Political Communication* 22, no.1 (January–March 2005): 45–62.

19. Jonathan Higgins, *Introduction to SNG and ENG Microwave* (Oxford, UK: Elsevier Focal Press, 2004).

20. Steven Livingston, "Diplomacy and Remote-Sensing Technology: Changing the Nature of Debate," *Net Diplomacy: 2015 and Beyond* no.16 (August 2002): 1–7; "The New Information Environment and Diplomacy," in *Cyber-Diplomacy in the 21st Century,* ed. Evan Potter (Toronto: McGill-Queens University Press, 2002), 110–27; "Remote Sensing Technology and the News Media," in *Commercial Observation Satellites: At the Leading Edge of Global Transparency,* ed. John Baker, Kevin O'Connell, and Ray Williamson (Santa Monica, CA: Rand Corporation and the American Society for Photogrammetry and Remote Sensing, 2001), 485–502.

21. For examples, see http://www.globalsecurity.org/eye/index.html.

22. And of course they still do, as John M. Hamilton and Emily Erickson point out in this volume. Parachute journalism, for better or worse, isn't going away. My point is that the first alert system, the initial detection and coverage of breaking news, is more likely to be done by amateurs wielding handheld devices of various types. David Perlmutter and Kaye Trammell make a similar point regarding bloggers.

23. Douglas Heingartner, "Honoring News Photos as Picture-Taking Evolves," *New York Times,* May 3, 2005.

24. Keith Richburg, interview by the author, Washington, DC, July 19, 2005.

25. Nicholas Wood, "Video of Serbs in Srebrenica Massacre Leads to Arrest," *New York Times,* June 3, 2005.

26. Ariana Eunjung Cha, "Do-It-Yourself Journalism Spreads; Web Sites Let People Take News into Their Own Hands," *Washington Post,* July 17, 2005.

27. W. Lance Bennett, "The Burglar Alarm That Just Keeps Ringing: A Response to Zaller," *Political Communication* 20, no. 2 (April– June 2003): 131–38. A fundamental premise of this essay is that the principal role of a free press, the very reason it is free, as in free of government or other limiting restraints, is that it serves the purposes of maintaining democratic rule. The press must be free to monitor malfeasance, corruption, and other forms of abuse by government and business. Of course, other theories of the press exist, though they are not applied here.

28. Philip Seib, *Going Live: Getting the News Right in a Real-Time, Online World* (Lanham, MD: Rowman and Littlefield, 2000).

29. Leon V. Sigal, *Reporters and Officials: The Organization and Politics of Newsmaking* (Lexington, MA: Heath, 1973), 121.

30. Fenton, 76.

31. Steven Livingston and W. Lance Bennett, Gatekeeping, Indexing, and Live-Event News: Is Technology Altering the Construction of News?" *Political Communication* 20, no. 4 (October–December 2003): 363–80.

32. Livingston and Van Belle.

33. Steven Livingston, "Media Coverage of the War: An Empirical Assessment," in *Kosovo and the Challenge of Humanitarian Intervention,* ed. Albrecht Schnabel and Ramesh Thakur (Tokyo: United Nations University Press, 2001), 360–84.

34. Lisa de Moraes, "Only CNN Gets the Picture," *Washington Post,* April 12, 2001.

35. For a classic description of the frustrations and anger that resulted from the necessity of relying on the U.S. military to transport video and other news to a satellite transmission facility during the Persian Gulf War, see John J. Fialka, *Hotel Warriors: Covering the Gulf War* (Washington, DC: Woodrow Wilson Center, 1992).

36. Philip Seib in this volume.

37. Quoted in Alvin Snyder, "Journalism: A Risky Profession," USC Center on Public Diplomacy Web site, July 21, 2005, http://www.ebu.ch/CMSimages/en/USC%20Center%20on%20Public%20Diplomacy%20_%20HEST_tcm6-43172.pdf.

38. Michael Schudson, *Discovering the News: A Social History of American Newspapers* (New York: Basic Books, 1978), 16.

39. Iyengar, 11.

40. Quoted in Iyengar, 28.

41. Iyengar, 27.

42. Ibid., 45.

43. Fenton, 18.

44. Data provided to author by BBC in London.

45. Meg James, "Networks Have an Ear for Spanish," *Los Angeles Times*, September 11, 2005.

46. Brian Ellsworth, "Venezuela Launches Cable News Station," NPR's Morning Edition, National Public Radio, July 18, 2005, http://www.npr.org/templates/story/story.php?storyId=4758465.

47. The question of real-time media's affect on foreign policy processes is referred to as "the CNN effect." See Steven Livingston and Todd Eachus, "Humanitarian Crises and U.S. Foreign Policy: Somalia and the CNN Effect Reconsidered," *Political Communication* 12, no. 4 (October-December 1995–96): 413–29.

48. Nicholas D. Kristof, "Terrorists in Cyberspace," *New York Times*, December 20, 2005.

49. Susan B. Glasser, "Probing Galaxies of Data for Nuggets: FBIS Is Overhauled and Rolled Out to Mine the Web's Open-Source Information Lode," *Washington Post*, November 25, 2005.

# 4

# BLOGGERS AS THE NEW "FOREIGN" FOREIGN CORRESPONDENTS

*Personal Publishing as Public Affairs*

KAYE SWEETSER TRAMMELL AND DAVID D. PERLMUTTER

## INTRODUCTION: BLOGGERS AS FOREIGN CORRESPONDENTS

As detailed in several other chapters in this book, the definition of "foreign correspondent" is increasingly confounded by new media technology and practices. Heretofore, the term evoked a picture of a ruffled "old China (or Moscow, or Paris, or Beirut) hand" whose baggage trail included a Remington typewriter, a whisky flask, and an expense account book from a major print publication or syndicate. In other words, the traditional source of foreign affairs coverage for most Americans was an American employee of a large American news organization. In contrast, consider the case of today's freelance reporter Christopher Allbritton, who successfully set up a personal Web site and asked readers to make online donations to finance his travels to Iraq.[1] The so-called PayPal journalist was then able to tell the story to his paid readers rather than adopt an editor's focus or bias. On his site, Allbritton calls his work "pure, individual journalism using a laptop and a satellite phone,"[2] done without the protection of the Department of Defense embedding program or a bulletproof vest.

But an even more striking revolution is under way in the technology, economics, and sociology of learning about foreign peoples, events, and issues via media. In the entire history of mass media and journalism, it was the norm for Americans to hear the voices of other peoples by listening to, reading, or watching media created and transmitted by the government or by the mainstream U.S.-based press. Foreigners—unless they were leaders, diplomats, or prominent dissidents—tended to be quoted only as human interest filler or exemplars of larger issues: e.g., "Ahmed the fisherman does not know what to make of the political upheaval in Cairo. . . ." The onset of the commercial Internet in the 1990s promised to bypass these traditional channels: people could access news of foreign lands via (a) the Web sites of their newspapers or governments, (b) independent media that in many cases challenged official points of view and mainstream media consensus, and (c) more rarely, the personal Web sites of the citizens of other countries. Then, beginning in the late 1990s and building to a tidal wave of popularity and power by 2004, a new genre of Web literature took front stage as a source of foreign affairs information. The phenomenon is the blog, short for *Weblog:* as most commonly defined, an online compendium of news, opinion, and debate by individuals.

Here we examine foreign independent blogs that have become extensions of, sources for, or replacements of traditional foreign affairs reporting. Their creators and contributors have the ability to talk to us directly, as the common phrase goes, "from ground zero," from Berlin to Beijing to the Congo. Perhaps more important, we the home-front audience can post comments, ask questions—that is, interact—with the stranger in a strange land, even if we are at war with her or his country. So to read a story about events in Baghdad, we no longer need CBS or the Associated Press; we have Salam Pax and his ordinary Iraqi's personal perspective describing the bombs falling outside his front door. Or to catch up on breaking news about the "Tulip Revolution" in Kyrgyzstan we can turn to an instantly created blog, Akaevu.net, and get the "scoop" while the traditional networks' correspondents are still on the plane from Moscow. In short, while in the past there have been times when the reporter becomes a part of the story,[3] this chapter explores how blogs allow the citizen reporter to become *the story.*

## BACKGROUND: THE WHOLE WORLD IS BLOGGING

Blogs, as of this writing in late 2006, are *the* hot media trend, although others like podcasting, YouTube, and Facebook offer new venues for the same developments. The Technorati Internet analysis service reports tracking 16 million blogs, from practically all nations.[4] Up to two-thirds of Americans on line report that they have consulted a blog some time in the last year. A recent Pew Internet and American Life report on the state of blogging found that many blog readers used the medium for political and campaign news during the 2004 U.S. presidential election.[5] Moreover, blogs have received an immense amount of media attention in connection with a number of blog-driven or blog-influenced events over the past few years, such as the resignation of Senator Trent Lott as Senate majority leader, the *60 Minutes* "Memogate," and the resignation of CNN's Eason Jordan.[6]

Blogs, however, are a young medium; the word *Weblog* was coined as recently as 1998. Many questions remain about blogs' actual numbers, their influence, and their future. Just as with the penny press of the 1840s, radio in the 1920s, and television in the 1950s, it would be foolhardy to predict that blogs, blogging, and the role of blogs will evolve in any particular direction. In terms of the relationship of blogs to foreign affairs reporting, policy making, and public opinion formation, however, several interesting trends and phenomena are clear.

### Blogs Are a Global Phenomenon

Blogs represent a giant leap in human communications technology, style, and reach. For the first time in history, any ordinary person can "mass communicate" to the Internet-connected planet. Projects like Global Voices, launched and sponsored by the Berkman Center for Internet and Society at Harvard Law School, have highlighted millions of blogs in various countries. In China, for example, there are an estimated seven hundred thousand bloggers. The third most commonly used language in the blog world is Persian. It should be cautioned that, at this time, the blog strata is wide but not yet deep. A vast majority of foreign bloggers are middle-class people or intellectuals who have computer skills: students, academics, businesspeople, professionals; as one of the authors of this chapter has noted elsewhere, "peasants don't blog."[7] Yet we can expect that, as it has in the United States, blogging will eventually percolate downward to reach broader socioeconomic classes. Some groups like the Soros Open Society foundation are attempting, in countries like Kyrgyzstan, to set up "village kiosks" where poor farmers can create their own blogs. The Global Voices project is attempting to link up foreign bloggers with American journalists—that is, to establish bloggers in, say, Lebanon, as sources.[8]

### Foreign Blogs Complement Traditional Foreign Affairs Coverage

As a sure sign that blogs are considered significant, many mainstream media organizations have "blogged up" their print, television, and Web presences. These measures include, as the organization CyberJournalist has documented, having on-staff correspondents create blogs.[9] In addition, it is now typical in any big foreign affairs story to "go to the blogs"—that is, for television news programs or print press reporters to mention "what the blogs are saying." Indeed, during the London bombings, the press regularly used blog-originated reporting (with pictures) from citizen journalists who were on the scene of the blasts and the aftermath. Britain's *Guardian,* for example, created an open blog on its Web site where anyone could post comments. On July 22, 2005, the paper's blog editor, Jane Perrone, introduced one reader's account of the second London bombing attack "on the Shepherd's Bush tube hit by one of the bomb attempts yesterday." The citizen journalist then reports:

> I was on the tube in Shepherd's Bush yesterday when the bomb went off in the next carriage. I have since been watching the news avidly and feel the need to give an eye witness account of what I experienced, as I feel there are some gaps in the media reports that should be filled.
>
> [. . .] As we went over the bridge, approaching Shepherd's Bush I heard what sounded like a small gunshot.
>
> Nobody reacted till about 30 seconds later when we noticed a disturbance in the next carriage. Suddenly we saw that smoke was filling the carriage and people were very distressed. [. . .]
>
> What I distinctly remember was a group of young teenage girls in long black headscarves and gowns, who had been next to the bomb, coming into our carriage they sat down and were rocking back and forth, I will remember the noise one of them made for the rest of my life, she was moaning out of pure horror and terror.
>
> What needs to be reported in the media is that the bomber had left a rucksack right next to a group of Muslim girls. If the bomb had gone off it was timed perfectly to go off while the train was on the bridge. We were completely stuck. Thankfully after about three minutes the train started moving and stopped in Shepherd's Bush, to let us off.[10]

Such "reporting" is more than human interest filler; it is a sign of people attempting, on their own, to fill in "some gaps in the media reports," both to create old-style media content and to supersede it. Mainstream media also turned to blogs to find out about commentary and opinions about the bombings, often quoting British and Arab bloggers. Such a phenomenon seems confirmation of the observation by a recent Pew report that bloggers serve "as a guide for the mainstream media to the rest of the Internet."[11]

## Blogs Offer Competition and Critique to Mainstream Reporting

Research shows that the more people use the Internet, the less likely they are to use traditional media, so one possibility is that increasing use of blogs for foreign affairs information gathering will correspondingly decrease consultation of television and print foreign news. It is also clear that, as younger and more Web-savvy people abandon traditional media, blogs represent a potential threat to "old media."[12] The May 2005 formation of the "Pajama Media" alliance between some of the most well-known political blogs aimed to create an "extensive network of globally affiliated blogs to provide first-person, in-depth coverage of most major news events, including both camera and video footage."[13] Indeed, the most popular blogs are the creations of individuals. Some of the major bloggers, like Andrew Sullivan, were media elites; others, like the creators of Instapundit and Little Green Footballs, had no connection to media or politics before "going blog." Blogs, then, allow anyone, within the purview of the webmaster, to comment, argue, rant, and hyperlink any other content on the Web or elsewhere. As exemplified by the eyewitness account quoted above, bloggers are not just supplementing the mainstream media but critiquing it. A large amount of blog discourse is devoted to critiques of old media, especially on foreign policy issues.

John Burgess, a retired foreign service officer, edits the blog "CrossRoadsArabia." Its specialization is the kingdom of Saudi Arabia. When, in late July 2005, King Fahd of Saudi Arabia died, Burgess dissected the problems with many media accounts that were inaccurate in either facts or analysis. For example, he writes:

> The International Herald Tribune has a piece on the new King. It repeats a lot of "conventional wisdom" about Abdullah that isn't actually right. Abdullah, for instance, has never been "anti-US." He

> just wasn't as pro-US as Fahd, displaying more of a tilt toward Saudi Arabia's well-being as his ultimate concern. That isn't how the US military took it, of course . . . anyone less inclined to simply sign a check and not ask questions was clearly going to be viewed less favorably. Perhaps for arms sales, but not for the stability of the kingdom. The article—bylined only "Associated Press"—is also a little high-handed in assuming that Sultan's being named Crown Prince equates to slowing down reform. While Abdullah was not bound by law to name Sultan as his successor, to not do so would have occasioned hot dispute within the family councils. The order of succession has been pretty clear in practice. Sultan's move up is entirely within keeping to that practice.[14]

Burgess has credentials to make such claims; lack of them, of course, does not prevent many other bloggers from attacking what is printed or telecast in regular media.

Equally, bloggers guard the guardians: they are an army of fact questioners, a vocation now moribund in many major media. As political scientists Daniel Drezner and Henry Farrell noted, in one case:

> In June 2003, the *Guardian* trumpeted a story in its online edition that misquoted Deputy U.S. Secretary of Defense Paul Wolfowitz as saying that the United States invaded Iraq in order to safeguard its oil supply. The quote began to wend its way through other media outlets worldwide, including Germany's *Die Welt.* In the ensuing hours, numerous bloggers led by Greg Djerijian's "Belgravia Dispatch" linked to the story and highlighted the error, prompting the *Guardian* to retract the story and apologize to its readers before publishing the story in its print version.[15]

Indeed, because bloggers are a diverse group—with many languages, backgrounds, interests—they are very likely to catch blunders, misquotes, and goofs that the monocultural journalist might not.

## Repressive Regimes Aggressively Seek to Censor Blogs

Blogging enthusiasts are often tempted to depict blogs as an unstoppable wave of human liberation, but new media technology can be used to enforce control as much as to promote freedom. The People's Republic of China, for example, has demanded that all its bloggers officially register with the government.[16] On a wider scale, China's giant state information ministries have actively worked to control Internet content.[17] These agencies scan the Web as well as personal and commercial e-mail for "heretical teachings or feudal superstitions" and any postings "harmful to the dignity or interests of the state," commonly referred to in state literature as "poisonous weeds."[18] Techniques for censorship include blocking certain Web sites from Chinese servers and filtering e-mail with search engines that seek out trigger words, ranging from the names of Chinese leaders to suspicious terms such as "freedom" and "democracy."[19] Other regimes, like that of Iran, have followed suit, and in many cases high-tech filtration is accompanied by old-fashioned coercion: throwing bloggers in jail or threatening them with fines.

## CASES: BLOGS FROM ABROAD

The blogs discussed here are those that have been seen as more than just journals; these are blogs that, while they may contain personalized viewpoints, report comprehensively on terrorist activity, strategic military bombing, and genocide—all occurring just outside the bloggers' front steps. They are not necessarily the "top blogs" of the regions they represent; rather, they exemplify changes in the sources, form, styles, and distribution channels through which American audiences (including journalists) receive information from foreign lands.

### Salam Pax's "Dear Raed"

In a study of the interplay of blogs with foreign affairs, Daniel W. Drezner and Henry Farrell noted,

> It was March 21, 2003—two days after the United States began its "shock and awe" campaign against Iraq—and the story dominating TV networks was the rumor (later proven false) that Saddam Hussein's infamous cousin, Ali Hassan al-Majid ("Chemical Ali"), had been killed in an airstrike. But, for thousands of other people around the world who switched on their computers rather than their television sets, the lead story was the sudden and worrisome disappearance of Salam Pax.[20]

Indeed, Pax—one young Iraqi's pseudonym—introduced the world public to viewing bloggers as foreign correspondents. His dispatches from Baghdad were seen daily leading up to and throughout the war. Because the blog was run on a free service, many questioned the genuineness of the so-called Baghdad Blogger, as blogs had already seen high-profile hoaxes.

Regardless, many felt that the realism and spirit in the writing proved the blog's authenticity. Beyond the major buzz created by the blog as it gained readers, journalists such as Peter Maass, a foreign correspondent writing for *Slate* magazine, praised the blog: "His [Pax's] lively and acerbic blog was far better than the stuff pumped out by the army of foreign correspondents in the country."[21] Maass was not the only traditional journalist to take note, as the *Guardian* observed: "It was the great irony of the war. While the world's leading newspapers and television networks poured millions of pounds into their coverage of the war in Iraq, it was the Internet musings of a witty young Iraqi living in a two-story house in a Baghdad suburb that scooped them all to deliver the most compelling description of life during the war."[22]

Despite this praise from the mainstream media and an international audience of readers on his blog, Salam Pax's agenda didn't begin with grand aspirations. The blog launched as many American blogs do: it was intended to keep friends (in this case, specifically a friend name Raed who rarely used e-mail and found blogs a more convenient way to keep in touch) up-to-date on Salam Pax's life. After the blogging began, Pax perceived that many Iraqi blogs focused on religion and that few were in English. At that point, Pax decided to identify himself as an Iraqi and—on the off chance that someone from outside his world found the blog—show what the real life of an Iraqi was like. Considering that there are only approximately fifty known blogs published from Iraq,[23] the celebrity of this blog carries even more weight.

Even after the war began and Salam Pax lost the technological means to "file" his stories with his growing list of readers, he kept writing. He wrote because "there will be excellent, amazing, very important stories to be told by lots of people. We, sitting in Baghdad in our protected four walls, were never going to be these stories. There are people who went through much more."[24]

From Web diarist to foreign correspondent, Salam Pax was able to engage and inform the world about what was happening at his doorstep through firsthand accounts. The entries mixed the tenets of journalism (who, what, when, where) with a unique perspective explaining the "why" as only someone whose home had become a war zone could. Furthermore, the humanity of the personal accounts and intermittent insertion of details (e.g., watching the movie *The American President* because Pax was sick of the news or cleaning up the house all day after a fierce sandstorm) created an environment in which readers could form parasocial relationships with the blogger. The rich details of his daily life experiences and feelings allowed readers to feel as if they personally knew Salam Pax or that he was their virtual friend.[25]

Posts from the blog received the most media attention when the *Guardian* began running excerpts, seemingly treating the posts as reports from a foreign correspondent. Eventually, the blogger who transformed the idea of foreign correspondence was himself transformed into a more traditional foreign correspondent, as he began writing for the *Guardian* and turned his blog posts into a book. This blog is noted as a turning point in the popularity of personal publishing and the political power blogs can wield by personalizing a military conflict thousands of miles away.

### View from Iran

Of all the Middle Eastern countries, Iran is said to be the one that has most firmly embraced blogs. In fact, with approximately 5 million online Iranians,[26] an estimated 75,000 blogs are published from the country. Some report Iran to have the third-highest number of bloggers worldwide. Many credit the seeming ubiquity of blogging to the youth-oriented demographic makeup of the country; indeed, 70 percent of Iranians are under thirty years old, too young to remember the Islamic Revolution in 1979.[27] Yet, the country has been labeled "the biggest prison for journalists" by Reporters without Borders.[28] Blogs have most notably been used during crisis situations in the country.[29] For example, during the student demonstrations in June 2003, blogs served as a primary source of information about what was happening, almost in real time.

Considering the geopolitical makeup of the country and the high price an outspoken individual exercising the power of personal publishing would pay, blogs represent an anonymous way to communicate to the outside world. As such, "View from Iran" serves as a cathartic release for the blogger and a key source of unfiltered information about the country. The style of the blog is one that offers a unique look into the blogger's daily life, experiences, and politics. The blog was originally maintained by two bloggers and is now updated weekly by only one.

"View from Iran" concentrates less on formal discussions of politics than on anecdotal descriptions of life and politics mixed together. This blog, like so many blogs that are purposely written in English (still the international language of blogging), seems targeted toward those unfamiliar with Iranian life. One post, for example, reads:

> Family life in Iran can be summed up simply: heaven for children, hell for teenagers and young adults, resignation for the middle-aged . . . Old? Depends.
>
> Children in Iran live in a particular children's heaven: they get sweets before dinner and are not put to be[d] early when guests arrive. They are always part of the conversation.[30]

Here the blogger is explaining the cultural differences in how children are treated in Iran. Along these lines, the post also makes a direct comparison with American culture and how Americans incorporate children in the family. However, at the end of the post, discussion turns more political and offers insight into how the country's political turmoil has shaped treatment of children:

> Iranians complain that the revolution has weakened their family bonds. "Brother cannot trust brother," Iranians tell you.
>
> K's [a friend of the blogger's] interpretation is a little different: "The revolution made people retreat into their homes where they got on each other's backs too much. They have nothing to do now except butt into each others' lives."[31]

### Radio Free Nepal

Radio Free Nepal is a group blog run anonymously so that contributors can express their experiences and opinions without fear of retribution. The blog, which is updated weekly, was started when King Gyandendra of Nepal issued a ban on independent news broadcasts and threatened to punish newspapers for reports that ran counter to the official monarchist line.

The early posts on the blog provide a flash introduction to the situation:

> At exact [*sic*] 10:00am, the King's address began. I had already known the government would be dismissed so wasn't surprised at all for that. But was not prepared for his takeover of the executive power. It was as for many people a bit surprising.[32]

The blogger goes on to lay out the ramifications of the situation, thereby explaining why the blog is so important:

> By the afternoon, it was clear that there is heavy censorship in the news of television and FM radios. I called a friend of mine at the Kantipur Television who told me that there are military personnel monitoring the news. There was a strict censorship.
>
> There was no flights coming and going. The airports were closed.
>
> [. . .] With all this, we moved into an "information isolation" because we have no access to information content neither we would be able to verify and report on any activities. We will be merely writing and publishing the materials that the state will avail us.[33]

Given the situation, it becomes clear that personal publishing under the protection of an anonymous Internet and filing posts using a pseudonym serve as a means not only to communicate with others within the country and share information but also to act as an embedded foreign correspondent in a country where the official news cannot be trusted. According to the journalist-turned-blogger who runs the site, one of the main goals is to raise international consciousness about the situation in Nepal in hopes of aiding the fight to bring democracy to the country.[34]

Because once-harsh editorial pages have been replaced with either empty white space or editorials about trees,[35] the government-seized media empire has lost its voice and credibility. Internet communication has been notably successful and a popular means with which to thwart such media clampdowns in China; the Nepalese case study further sends the signal that it is impossible to quiet the voice of the repressed in today's Internet age.

## Slugger O'Toole

Voted the "Best Political Weblog" in the 2005 European Weblog Awards, Slugger O'Toole was conceived and is maintained by Mick Fealty. Fealty, who lives in England, began the blog as a personal site to catalog his research on Northern Ireland politics. This blog differs from some of the other media discussed here because it does not focus on firsthand accounts; rather, by sharing news articles and short commentary, it sparks debate on everything from political elections to Irish culture and language.

The blog attracts readers from around the world, but more importantly it has become a forum for people with differing political views. Fealty suggests that the opposing opinions introduced and discussed within the blog often push political conversations beyond the surface.[36] Furthermore, he asserts that the exchange of ideas from people on both sides of an issue can lead to the revelation of "insider information." In addition to the exchange of ideas across party lines, this blog has been said to catch the attention of politicians who keep up with the online discussions.[37]

The posts on Slugger O'Toole are short, conversational, and nearly always point outward to other sites or media articles. In a way, Fealty is quickly sharing something he came across with his readers then moving on with his day. Even if Fealty's posts seem unimportant, nearly every post on the blog has comments, with some posts inciting far more discussion than others.

Like other analyses of comments on political blog posts, the reader comments for Slugger O'Toole often refer to other comments rather than the original post.[38] Even so, the comments left in response to a Slugger O'Toole post stay on topic more frequently than those of some other blogs.[39]

## Akaevu.net

On March 24, 2005, President Askar Akayev, strongman ruler of the former Asian Soviet Republic of Kyrgyzstan, fled the country after a series of public protests. The so-called Tulip Revolution (referring to the rare mountain flower brandished by the demonstrators) was variously dubbed a "garden-variety" coup,[40] a "scary democratic rebellion,"[41] and even a CIA black op.[42] Progovernment Russian media labeled the events in Kyrgyzstan a U.S.-backed coup[43] and unconstitutional.[44] Media in the neighboring Central Asian republics either ignored the events or condemned them. But in the days of the unrest and the coup, few foreign news media workers or organizations were in any position to say much, since they had no reporters on the scene. Local Kyrgyz were even less informative, since almost all press outlets were controlled or owned outright by Akayev and his family or allies. State-controlled media, confused as to who was in charge, produced unreliable and erroneous accounts,[45] which forced many people to search for alternative sources of information, often Internet-based.

Enter Akaevu.net ("Akaevu net" in Russian means "Down with Akayev"), an advocacy blog that was created by the author of several other previous opposition Web sites.[46] At first glance, the blog would seem to be the loneliest form of opposition in a Central Asian republic where computers with Internet access available to the poor rural population can be counted in the single digits.[47] But in revolutions, raw numbers may not be decisive. A few Bolsheviks, for example, were able to seize Russia in 1917, whereas millions of protestors could not move the Chinese government in 1989. Indeed, in Kyrgyzstan only about a thousand demonstrators actually took over the government building and sent the president packing.[48]

The Web site's editor, Ulan Melisbek, a Kyrgyz citizen currently residing in the United States, described Akaevu.net as follows:

> As a result of the foul order by the Akayev-Toigonbaev [Akayev's son-in-law] gang, the most popular sites of Kyrgyzstan, Gazeta.kg and Kyrgyz.us, have been blocked. Access is also blocked to the popular regional resource Centrasia.ru, which is also covering the events in our country. Our response to Chamberlains-Akayevs will be the creation of innumerable sites on various servers, so that they shake up the financial position of Toigonbaev. Hackers are people who value their time and skills, and sooner or later Toigonbaev will become weary of paying for blockage of numerous sites.

During the protest, the site provided:

- from-the-scene news updates by citizen journalists
- pictures from digital cameras and cell phones

- innovative rolling, interactive pubic opinion surveys on topics such as "Should Akayev be impeached or given the status of First President with all privileges?" "Who should be the next president of the Kyrgyz Republic?" "What should we do with the Akayevs?" and "Should force be used to calm down Osh and Jalal-Abad?"[49]
- editorials and commentary by opposition figures and ordinary bloggers
- roundups of foreign press reaction (for those in the country who could not get access to those Web sites)
- roundups of analysis by nongovernmental organizations

In sum, Akaevu.net provided an alternative voice, indeed at times the *only* voice in a chaotic foreign affairs event.

## CONCLUSION

In looking at the blogs and corresponding phenomena described here, it is clear that untrained (or "citizen") journalists are coming to the Internet, blogging, and adding to the information and understanding about certain world events. What, then, is the mark of success? Must these blogs be read by international leaders and then spur policy changes? Is it enough that these blogs are perused by the world's citizens and that those who read them are touched? Such at least might be a baseline for weighing the alleged powers of the foreign blogger. As stated in the "manifesto" of the Harvard-based Global Voices project, "We believe in the power of direct connection. The bond between individuals from different worlds is personal, political and powerful. We believe conversation across boundaries is essential to a future that is free, fair, prosperous and sustainable—for all citizens of this planet."[50] Blogs enable all of us—if we have access to them despite economic or political obstacles—to engage in such "conversations" for our own edification and, we believe, to the enrichment of both democracy and foreign affairs coverage.

But in weighing the "power" of the press, simple dialogue is often not considered enough added value. Let us examine the Salam Pax case as a benchmark for measuring the "success" of a blogger turned foreign correspondent. Was U.S. president George W. Bush or his generals who were running the war reading Pax's blog as the military campaign evolved?

Probably not. Did the blog unequivocally result in changes in foreign policy or presence in Iraq? No. Yet the blog—and others discussed here—are seen as successful ventures in publicizing and at times reporting news from abroad. We argue here that these blogs provided a human-scale view of important world events. No longer is the enemy or the oppressed a faceless foreigner: he is someone with a family, someone we can receive updates from every day, someone the reader grows to care for and worry about.

Blogs may also develop into agents of what might be called citizen diplomacy. This phenomenon was not unknown in foreign affairs. Famously, for example, President Nixon's landmark trip to China in 1972 was paved by the travels of an American table tennis team in 1971. The message of simply seeing ordinary people "getting along" is a powerful one in any context and any age. Some preliminary research on blogging suggests that extra-national communities can be created. One study found:

> Bloggers on the whole perceive a shared sense of community in the blogosphere. Notwithstanding the social, political, and economic differences between the regional cultures of our participants, bloggers painted a remarkable picture of congruity in their experiences with activism, reputation, social connectedness and identity. Thus, we can posit that bloggers themselves represent a unique culture that permeates through regional boundaries.[51]

Such linkages may assist the easing of international conflicts at the all-important street level. We may ask, for instance, what happens when a critical number of Israeli and Palestinian bloggers contact each other? As we have seen, the outcome may be flame wars and partisanship, but it could equally be a humanizing and contextual understanding less likely to occur as a result of coverage from national media.

In sum, we do not argue that foreign blogs will replace foreign correspondents, but we predict that they will enrich the information the world receives about important events. Specifically, these personalized posts dispatched from the front lines provide an insight and tell a story that an American deployed to the event cannot. Citizens abroad and policy makers alike can read the accounts on blogs from within a nation and expose themselves to alternative viewpoints. Journalists can broaden their sources. Furthermore, the world can get a glimpse into a community and gauge its way of thinking. The one-sided view of a foreigner traveling across the globe to tell a story can be improved by tapping into these bloggers serving as foreign correspondents in their own land.

That said, the foreign blog is ripe for further examination; news researchers and professionals would be unwise to ignore their future developments. Scholars interested in the news gathering and gatekeeping process should examine the differences between traditional foreign correspondents and bloggers through in-depth analyses of the process each goes through in producing a news item for publication. Questions about source credibility and information quality should be posed. Finally, scholars should investigate the amount of attention these personalized reports are getting at various levels of government. Are policy makers reading these blogs? Are governments monitoring them and considering information published in them as a form of intelligence? Will these blogs ever have a direct impact on foreign policy or public opinion? The future is unwritten, but it will likely be blogged.

NOTES

1. Mark Baard, "Reporter Takes His Weblog to War," *Wired,* http://www.wired.com/news/conflict/0,2100,58043,00.html.

2. Christopher Allbritton, "Resume," Back to Iraq Weblog, http:// www.back-to-iraq.com/personal/resume.html.

3. Andrew Paul Williams, "Media Narcissism and Self-Reflexive Reporting: Metacommunication in Televised News Broadcasts and Web Coverage of Operation Iraqi Freedom" (Ph.D. diss., University of Florida, 2004).

4. Technorati, "About Us," Technorati Web site, 2005, http://www.technorati.com/about/.

5. Lee Rainie, "The State of Blogging," Pew Internet and American Life Project Web site, http://www.pewInternet.org/PPF/r/144/report_display.asp.

6. Howard Kurtz, "In the Blogosphere, Lightning Strikes Thrice," *Washington Post,* February 13, 2003.

7. David D. Perlmutter, *Blogwars: The New American Political Battleground* (New York: Oxford University Press, forthcoming).

8. David D. Perlmutter, "Will Blogs Go Bust?" *Editor and Publisher,* August 4, 2005, http://www.editorandpublisher.com/eandp/article_brief/eandp/1/1001009362.

9. The *Washington Post,* for example, has dozens of blogs (http://blog.washingtonpost.com) by existing staff who blog in addition to their regular writings and by others who are hired exclusively to blog.

10. Jane Perrone, "Attack on London 11:18 a.m.," *Guardian* Web site, http://blogs.guardian.co.uk/news/archives/2005/07/22/she_was_moaning_out_of_pure_horror_and_terror.html.

11. Michael Cornfeld et al., "Buzz, Blogs, and Beyond: The Internet and the National Discourse in the Fall of 2004," Pew Internet and American Life Project Web site, May 16, 2005, http://www.pewinternet.org/ppt/buzz_blogs_beyond_Final05-16-05.pdf.

12. Dan Gillmor, *We the Media: Grassroots Journalism by the People, for the People* (Sebastopol, CA: O'Reilly Media, 2004).

13. Roderick Boyd, "Three Political Web Logs Make a Run for the Mainstream," *New York Sun,* May 3, 2005, http://www.nysun.com/article/13179.

14. John Burgess, "King Fahd Dies, Abdullah Succeeds," *CrossroadsArabia,* http://xrdarabia.org/blog/archives/2005/08/01/king-fahd-dies-abdullah-succeeds/.

15. Daniel W. Drezner and Henry Farrell, "Web of Influence," *Foreign Policy,* November–December 2004, 35.

16. "Chinese Blogs Face Restrictions Tuesday," BBC News Web site, June 7, 2005, http://news.bbc.co.uk/2/hi/technology/4617657.stm.

17. Christopher Cox, "Establishing Global Internet Freedom: Tear Down This Firewall," in *Who Rules the Net? Internet Governance and Jurisdiction,* ed. Adam Thierer and Clyde Wayne Crews Jr. (Washington, DC: Cato Institute, 2003), 3–12.

18. Xiao Qiang, "Chinese Whispers," *New Scientist,* November 27, 2004, http://technology.newscientist.com/channel/tech/mg18424755.500-chinese-whispers-.html

19. Jonathan Zittrain and Benjamin Edelman, "Empirical Analysis of Internet Filtering in China," Harvard Law School, 2003, http://cyber.law.harvard.edu/filtering/china/.

20. Drezner and Farrell, 32.

21. Peter Maass, "Salam Pax Is Real," *Slate,* June 2, 2003, http://slate.msn.com/id/2083847/.

22. "Salam's Story," *Guardian* Web site, May 30, 2003, http://www.guardian.co.uk/Iraq/Story/0,2763,966819,00.html.

23. Claude Salhani, "Politics and Policies: The Other Mideast Revolt," United Press International, March 2, 2005.

24. "Salam's Story."

25. Kaye D. Trammell, "Celebrity Blogs: Investigation in the Persuasive Nature of Two-Way Communication Regarding Politics" (Ph.D. diss., University of Florida, 2004).

26. Jason Motlagh, "Words Are Weapons for Iranian Bloggers," United Press International, February 17, 2005, http://washtimes.com/upi-breaking/20050214-050322-8970r.htm, accessed March 5, 2005.

27. Ibid.

28. Ibid.

29. Ibid.

30. E.T., "About Families for San Diegan," View from Iran Weblog, May 27, 2005, http://viewfromiran.blogspot.com/2005/05/about-families-for-san-diegan-i-have.html. The blog describes itself as "Chronicling the adventures of an American in Tehran with guest appearances from her Iranian husband" and thus is a sort of amphibian blog, trying to cross over international boundaries for the reader.

31. Ibid.

32. Kathmandu, "The Day Log," Radio Free Nepal Weblog, February 2, 2005, http://freenepal.blogspot.com/2005/02/day-log.html.

33. Ibid.

34. Mark Glaser, "Nepalese Bloggers, Journalists Defy Media Clampdown by King," *Online Journalism Review,* February 23, 2005, http://www.ojr.org/ojr/stories/050223glaser/; Alan Connor, "Not Just Critics," *BBC News Magazine,* June 20, 2005, http://newswww.bbc.net.uk/1/hi/magazine/4111330.stm.

35. Glaser.

36. Mark Devenport, "Politicians Monitor the 'Bloggers,'" *BBC News* Web site, February 26, 2004, http://news.bbc.co.uk/1/hi/northern_ireland/3490568.stm.

37. Ibid.

38. Trammell, "Celebrity Blogs."

39. Kaye D. Trammell, "Looking at the Pieces to Understand the Whole: An Analysis of Blog Posts, Comments, and Trackbacks," paper presented at annual meeting of International Communication Association, New York City, 2005.

40. This section of the chapter is drawn from Svetlana V. Kulikova and David D. Perlmutter, "Blogging Down the Dictator? The Kyrgyz Revolution and Samizdat Websites," *International Communication Gazette* 69, no. 1 (2007): 29–50. See also: C. S. Smith, "Kyrgyzstan's Shining Hour Ticks Away and Turns Out to Be a Plain, Old Coup," *New York Times,* April 3, 2005; E. Burkett, "Democracy Falls on Barren Ground," *New York Times,* March 29, 2005, http://www.nytimes.com/2005/03/29/opinion/29burkett.html.

41. E. Sullivan, "A Scary Democratic Rebellion in Kyrgyzstan," *Cleveland Plain Dealer,* March 27, 2005.

42. R. Spencer, "American Helped Plant Tulip Uprising," *Ottawa Citizen,* April 2, 2005; John Laughland, "The Mythology of People Power," *Guardian,* April 1, 2005.

43. Y. Yuferovam, "Kirgizskiy perevorot" [The Kyrgyz Coup], *Rossiyskaya Gazeta,* March 30, 2005. Translation provided by Svetlana Kulikova.

44. M. Leontiev, "Kirgizskiy zvonok dlya Rossii" [Kyrgyzstan Rings the Bell for Russia], *Komsomolskaya Pravda,* March 28, 2005. Translation provided by Svetlana Kulikova.

45. For example, the national news agency Kabar during the day reported or cited other sources that Akayev was in the country at his residence, then later allegedly went to Kazakhstan, then to Russia, and finally admitted not knowing the president's whereabouts.

46. The blog is available but inactive since April 25, 2005.

47. Daniela V. Dimitrova and Richard Beilock, "Where Freedom Matters: Internet Adoption Among the Former socialist Countries," *Gazette: The International Journal for Communication Studies* 67, no.2 (2005): 173–87.

48. "Kyrgyz President's Palace Overrun," BBC News Web site, March 24, 2005.

49. Cities in Kyrgyzstan where unrest took place.

50. "Global Voices Draft Manifesto," http://cyber.law.harvard.edu/globalvoices/wiki/index.php/Global_Voices_Draft_Manifesto.

51. Norman M. Su, "A Bosom Buddy Afar Brings a Distant Land Near: Are Bloggers a Global Community?" paper, Second International Conference on Communities and Technologies (C&T 2005).

# 5

# U.S. MEDIA TEACH NEGATIVE AND FLAWED BELIEFS ABOUT AMERICANS TO YOUTHS IN TWELVE COUNTRIES

*Implications for Future Foreign Affairs*

MARGARET H. DEFLEUR

The emphasis of many of the chapters in this book is on foreign affairs, the journalists who cover it, and how coverage of world events shapes American public opinion about foreign affairs. In other words, we are discussing how what happens "out there" (in other nations) gets reported to us here in the United States and how this shapes our beliefs about the importance of foreign affairs and foreign policy.

However, it is equally important to consider the opposite perspective: How do the events that happen here get to them "out there" (in other nations), and how does this shape their beliefs and attitudes about us? This, too, may have an impact on foreign affairs and foreign policy.

Since September 11, 2001, much of our foreign affairs reporting has been concerned with the question of how Americans are viewed abroad and why others abroad often hold negative opinions and attitudes toward the United States and its people, leaders, and policies. However, the role of media content produced in the United States and exported globally, undoubtedly an important factor, has been little examined.

Thus the research reported in this chapter focuses on a different topic from others in this book. It is a study of one possible consequence of the flow of mass communications *out of the United States to other parts of the world.* It assesses potential influences on foreign affairs that can result from using new and old technologies to export, over many decades, mass media content to audiences of hundreds of millions in other countries. Such media content has long been produced in the United States and disseminated to worldwide media markets. Such products include not only news stories about people who live in the United States, but also motion pictures and television entertainment programs, as well as other forms of popular culture, that incidentally depict what Americans are like. These mass communication products portray actions that may be interpreted abroad as *typical* of the conduct, values, and characteristics of ordinary Americans.

From such sources, people in other countries may learn (what they believe to be) the nature, values, and behavior of the American population. These understandings, beliefs, and mental images, along with what is obtained from news about actions of American people, and of the U.S. government, provide a rich foundation for perceptions and interpretations of what Americans are like. Those interpretations may have significant influences on the nation's foreign affairs.

## BACKGROUND

The history of both news and entertainment received in countries around the globe is closely linked to the adoption of innovation in communication technologies. That process began long ago and has been taking place for a very long time. As early communication technologies were invented, developed, and adopted worldwide, an outward flow of news and media entertainment from the United States to world markets began. As additional technologies came into being, that flow intensified.

### The Long-Term Development of Communications Technology

Stories about people and events in the United States have been swiftly exported to audiences in other nations for over a century and a half—beginning with the successful completion in 1866 of a functioning transatlantic cable. That telegraphic means for transmitting news accounts originating in the United States to European countries was soon followed by a network of such cables linking most parts of the globe. By that means, people in most literate countries were able to learn from their newspapers about people and events in the United States, even during the 1800s. For example, news about the U.S. Civil War was widely reported. The adventurous reporter "Nellie Bly" (Elizabeth Cochrane) published a scandalous exposé of the treatment of mentally ill persons at Blackwell's Island, a mental institution in New York City. Her report of the conditions at the hospital received worldwide attention.[1]

By the end of the nineteenth century, the technology of motion pictures had become a reality, and a huge new and very popular entertainment industry soon came into existence in both Europe and the United States. However, as World War I broke out in 1914, the fledgling movie industry in Europe was virtually shut down. That left world markets to be supplied by the products of Hollywood, which because of its weather had become the locus of American movie-making.[2] Even today, films made in the United States dominate world markets.

Scheduled radio broadcasts of both news and entertainment, received on sets at home, began in both the United States and Europe in the early 1920s. Similar broadcasts soon started in other parts of the world. It was a medium ideally suited for the transmission of popular culture. Popular music—jazz, swing, rock, and so forth, flowed from American performers and musicians via records and radio. Young people around the world became a major audience for the latest popular songs and their celebrity performers, brought to them by mass communications.

During the 1950s and 1960s, television was rapidly adopted, and TV programming became widely available to people in their homes. Few countries had their own production facilities, so recorded content was imported and rebroadcast on local systems. In time, TV programs and especially movies, on recorded tape or DVDs, were also readily available, sometimes in foreign countries before Americans could obtain them. Satellite dishes permitted reception of broadcasts where cable was not available. Even more recently, as personal desktop and laptop computers have come into wide use, the World Wide Web has been added to the mix. Today, cell phones with picture reception capacity can receive mass communication transmissions.

As these various media technologies spread around the world, they provided the means by which American-based multinational corporations could produce and export mass communication content to audiences in countries around the globe. Because few or no facilities were in place in some countries for generating their own news and entertainment for their populations, it was far easier to show American films, transmit via radio the latest news, broadcast popular music, and fill hours of television with entertainment programming imported mainly from the United States.

Thus successive generations in other nations who have received this flow of mass communications for well over a century have been learning about and developing images of the people who live in the United States.

Today, a limited number of highly competitive multinational corporations continue to produce such news and entertainment products and disseminate them for profit throughout the world. As a consequence, as these products are provided to populations via local media, young people in other countries get to "know" what ordinary people who live in communities in the United States are like. They read about them in their newspapers and see them with their own eyes on movie screens and on TV sets. For those who have never traveled to the United States or actually known an American, what they encounter in their media may be a primary source of learning about the people who live in the United States. The beliefs and images developed from such media sources may have unwanted consequences for the nation's foreign affairs.

Clearly, there are many other possible sources of information about Americans. Other significant sources of influence include families, peers, teachers, religious leaders, and government officials, to name a few. However, as data presented in this chapter will indicate, there is also reason to conclude that the depictions of Americans, embedded in the massive flow of media content, are a dominant source of information for many who attend to them.

## News and Entertainment Production for Contemporary Global Markets

What many audiences around the world see when attending to mass communications is much the same as people see in a typical community in the United States—that is, depictions of American people as their newsworthy activities are reported by the press and in portrayals of what they are like in the plots and dramas of films, reality shows, soap operas, police dramas, and other forms of popular culture.

The news about Americans and the United States that is sent to international audiences may often be the same that is transmitted to domestic audiences. There may be no deliberate attempt on the part of those who develop and disseminate this news to have it serve as a means of providing objective lessons about the behavior or values of the majority of persons who live in the United States, yet these unintended lessons may have serious consequences. An unfortunate example of the consequences of the media's negative portrayal of Americans took place in May of 2005, when *Newsweek* reported that a copy of the Koran had been flushed down a toilet by American personnel in the prison at Guantanamo Bay.[3] Although the magazine later acknowledged errors in the story, the episode provided an unintended lesson concerning the lack of respect that some Americans hold for the Muslim faith.

The design, development, and transmission of entertainment content that the media provide are also of concern. Movies, soap operas, "reality" shows, and cop and crook dramas are produced and shown to audiences, not as a public service but for profit. These productions are then exported widely to other nations for broadcasting. Street vendors sell or rent recorded versions of movies and TV programs cheaply, so in places where broadcasts or cable are not available, the content is still readily available. Satellite receivers provide another means of reception. Thus American entertainment and news programs are accessible in all but the most remote parts of the globe. In that sense, as audiences view what Americans are like in these programs, the media may serve as *weapons of mass instruction,* often presenting distorted and negative images of the nature, values, and behavior of the people who live in the United States.

As the findings reported in this chapter indicate, there is little doubt that the content provided by such mass communication is attended to both widely and enthusiastically. In particular, media entertainment—especially movies and TV programs—is enjoyed by many millions in other countries in much the same way as in the United States. It is the young—the oncoming next generations who are not yet preoccupied with the responsibilities of adulthood—who are the heaviest consumers of such popular culture presented by the mass media. As each generation moves into adulthood, its members will take over government, businesses, and other responsibilities in their countries, and therefore the ways in which imported mass communications have prepared them to think about Americans should be an important consideration in U.S. foreign affairs. While many other factors and events influence beliefs and attitudes toward Americans, the question addressed in this chapter is, given the depictions and portrayals of Americans in the mass communications that young people in other countries receive, what beliefs and images do they develop about the activities, actions, and values of people who live in the United States?

## ASSESSING TEENAGERS' IMAGES OF AMERICANS IN TWELVE COUNTRIES

To determine the ways that young people in three regions of the world (Muslim, Asian, and Western) think about ordinary Americans, DeFleur and DeFleur in a recent large-scale study assessed the beliefs and attitudes toward Americans of high school–age youths.[4] This was a different research objective than the more common one of assessing the beliefs of adults in other countries concerning official U.S. policies, leaders, military actions, or other "official" aspects of the United States.

A total of 1,313 high school teenagers between fourteen and nineteen years of age (the average was seventeen) responded to a questionnaire administered to them in their classes by local teachers or high school administrators. The questionnaire also assessed the students' contact with Americans, their uses and preferences of media, and especially their exposure to mass communications produced in the United States.

The questionnaire was carefully translated into local languages and checked for understandability by participating school teachers and professional translators. The questionnaire included a number of items assessing the youths' views of *ordinary Americans*—those typically seen at work, in family settings, at the supermarket, as neighbors, and so on. The countries included in the study were Saudi Arabia, Bahrain, South Korea, Mexico, China, Spain, Taiwan, Lebanon, Pakistan, Nigeria, Italy, and Argentina.

A major goal of the project was to document the nature of the teenagers' beliefs and attitudes toward ordinary people who live in the United States. However, the researchers also hoped to provide explanations of how such youthful views are shaped by globally distributed media products, as well as possible long-term consequences of exposure to these influences. The central focus of the present chapter is on the role of mass communications in shaping beliefs and attitudes of young people—the next generation who will soon be adults—and to assess the implications of the findings for future American foreign affairs.

Of course, the study is limited in a number of ways. First, the researchers used a nonrandom, purposive sample of students for the study.

A network of foreign graduate students at a large northeastern U.S. university volunteered to assist with the study. The graduate student volunteers, along with their families, had direct relationships with school administrators and teachers in their home countries. The graduate students contacted school principals for permission to present the questionnaires and explained the purpose of the study. They asked the teachers if they would be willing to administer the questionnaires to the students in their schools. Consequently, all questionnaires were personally administered in classrooms by the students' teachers.

Although this strategy enabled the researchers to gather data from hard-to-reach populations, it had obvious sampling limitations. In addition, the researchers were not present at the schools to verify that all instructions and procedures were followed. As a result, the findings from the study cannot be generalized to all high school students in the nations studied. At the same time, however, the data gathering strategy had a significant advantage. Those who administered the questionnaires were persons who were familiar to the students and trusted by them. This bond of trust was especially important in those situations where students were more likely to be suspicious of outsiders or distrustful of someone from the United States. All of the students and their teachers were given guarantees of anonymity by the graduate students and the researchers.

## LESSONS ABOUT AMERICANS IN MEDIA ENTERTAINMENT

The present chapter is focused mainly on the content and consequences of exported popular culture such as TV programs, motion pictures, music videos, and recordings, as well as other entertainment products. Undoubtedly, when depicting Americans in such content, the producers are not deliberately trying to influence anyone's beliefs or attitudes toward Americans. Their goal is not to produce instructional materials about Americans but to make money. To maximize profits their products are designed to appeal to the largest segment of the market within the populations of the many countries to which their wares are exported. That segment of the market is the young. Figure 1 shows the age distribution of the "developed" (mainly industrialized) and the "developing" (essentially nonindustrialized) countries in the world. It is apparent that the young are far more numerous than those who are older.

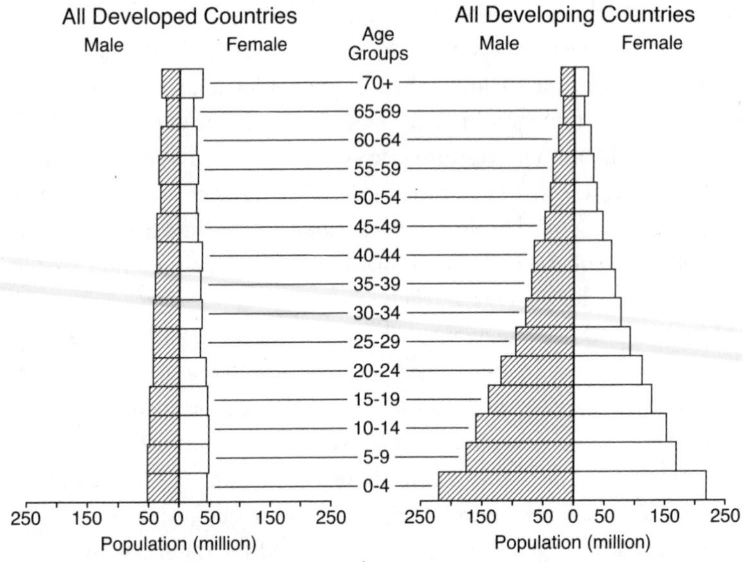


**Figure 1.** Population pyramids
Based on data from UN Department of Economic and Social Affairs, Population Division, "World Population by Gender" (New York: United Nations, 2003).

News and entertainment producers understand this distribution, and they understand that the tastes and interests of the young are different from those of their elders, who are less numerous and more traditional. Therefore, the market wherein the profits lie is among people who are in their teens or early twenties. In conservative societies, in which the behavior of young people is closely controlled by their elders, teenagers may conform to local limits on their conduct. However, imported media entertainment can allow them to enjoy vicariously the nonconforming behavior that is portrayed as typical of young Americans. In many cases, such behavior would be discouraged, or even forbidden and punished, within the norms governing their own lives. For that reason, even media depictions of children disobeying their parents, of women wearing revealing clothing, of people holding hands or flirting in public can seem exciting. Students from the conservative societies studied in the project have confirmed this observation.[5] Seeing depictions of young Americans doing what they do on TV and movie screens, activities which are often rigidly forbidden in some societies, can be fascinating.

Above all, young people the world over want to see exciting action. They enjoy viewing explosions, fistfights, car chases, shootouts, criminal activities, and other portrayals of dangerous behavior on TV and movie screens. Many are also attracted to scenes of nude women and depictions of couples engaging in sexual behavior.

Exactly this kind of exciting content has been increasingly incorporated into the motion pictures, TV dramas, music videos, and other forms of popular culture that are produced by media corporations and shipped to other countries. The reason for continuously pushing the lines of showing unacceptable conduct is that competition for audiences among those who produce media entertainment products is fierce. The products must attract the attention of the audience in order to recover the costs of production. Increasing the excitement level in what is produced enlarges that audience and thereby enhances profits.

The consequence of this situation is that these forms of media entertainment provide subtle but abundant incidental lessons to young people in societies around the world about the values and behavior of ordinary people who live in the United States as they are portrayed in movies and depicted daily in television programs such as *Baywatch, Desperate Housewives, The Sopranos,* and *Sex and the City.*[6] The lessons that they learn may be unplanned by those who produce such entertainment products, but they are learned unwittingly by the audience as it enjoys what this type of media entertainment provides.

## OVERALL ATTITUDES TOWARD AMERICANS

The questionnaire to which the 1,313 teenagers responded contained a total of twelve statements about Americans. The statements represent evaluative beliefs about characteristics of Americans held by the respondents. They could select one of the following response categories as the one best expressing their beliefs about each statement: "Strongly Disagree," "Disagree," "I am Undecided," "Agree," or "Strongly Agree." (This is a standard strategy for assessing how people feel about an issue by using a Likert-type questionnaire.) A numerical scale was developed to represent these categories, in which a –5 represented the strongest level of disagreement, zero was a neutral point, and +5 indicated the strongest level of agreement. The teenagers were also asked for other information, such as whether they or their families had traveled to the United States, whether they had had contact with American citizens, and whether various media were available and used in their households. They were also asked about their television viewing and movie attendance, and demographic information such as age and gender was collected.

In addition, the high school students often wrote lengthy comments, sometimes filling an entire page, on their feelings about Americans. Some of the students used very explicit language to describe Americans. A few drew pictures to aid in understanding their comments. Others thanked the researchers for asking their opinions and added that they were sorry to respond with negative comments.

A very negative picture emerged from the study. Briefly stated, the findings indicate that among the majority of the youths studied, Americans are thought to be *violent* toward each other, to be *overly materialistic,* to have *little respect* for people unlike themselves, to want to *dominate* others, and to engage often in *criminal activities.* American women, in particular, are thought to be extremely *sexually immoral,* especially when judged against local standards of female conduct. Indeed, those studied saw little for which to admire Americans.

While it may not be surprising that others think negatively of Americans, what is surprising is the extent of those negative beliefs on the part of the students who were studied. The teenagers made very few positive comments.

How did the teenagers in the countries studied acquire such negative pictures in their heads? It was not by personal experience with actual Americans. Few of those studied had ever known an American or had ever traveled to the United States. (Those who had were largely from Mexico.) What this indicates is that one of their principal sources of knowledge about Americans was what they saw reported in the news and how Americans were depicted in movies and on TV.

As Figure 2 indicates, the students' attitudes ranged from decidedly negative on the part of teenagers in Muslim countries in the Middle East, such as Saudi Arabia, Bahrain, and Lebanon, through somewhat negative in others, with neutral attitudes found only in Italy, and slightly positive attitudes in Argentina. On average, however, the teenagers studied overall had negative attitudes toward the people who live in the United States.

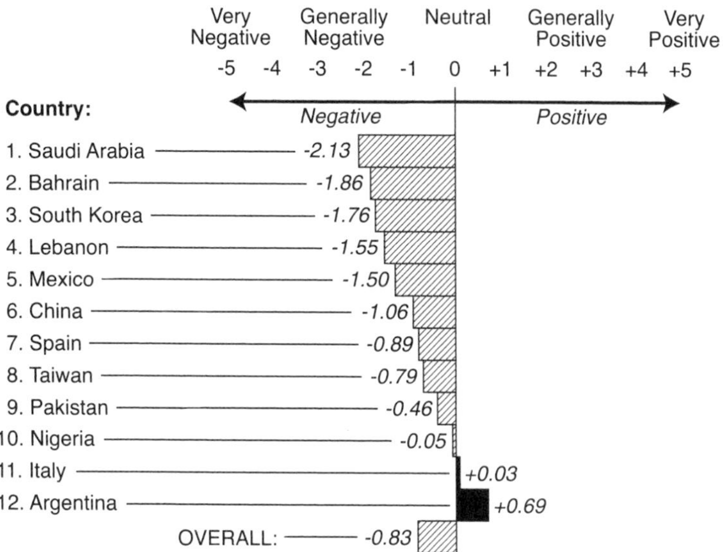


**Figure 2.** Overall attitudes toward Americans by country

It is interesting to note that the teenage respondents in South Korea and Mexico, nations considered allies and partners of the United States, held attitudes toward Americans that were almost as negative as those in Saudi Arabia and Bahrain. Obviously, many factors play a role in shaping those attitudes, but from the written comments of the respondents, it seems clear that media content is one important factor.

## Specific Beliefs about Americans

Figure 3 shows the specific statements to which the teenagers responded and indicates the degree to which the teenagers from all twelve of the countries combined expressed positive or negative feelings about these statements. Overall, the teenagers strongly believed that Americans are violent, immoral, materialistic, and domineering. They also believed that Americans are not generous or peaceful, have little respect for others, and engage in criminal activities. Indeed, there was little for which they admired Americans.

Of course, the pattern of positive and negative responses differs from one nation to the next. For example, the combined responses of teenagers

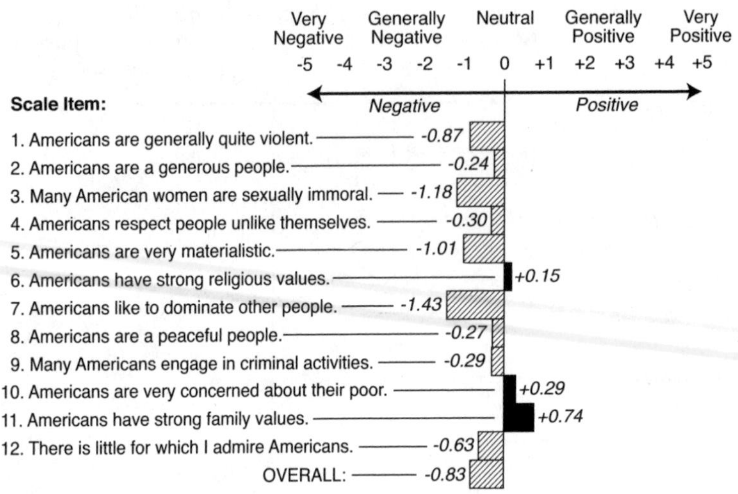


**Figure 3.** Overall attitudes toward Americans by scale item

in Saudi Arabia, Bahrain, and Mexico are all negative. The teenagers in China and Spain believe that Americans have strong religious and family values. The teenagers in Nigeria believe Americans are generous and concerned about the poor. The Italian teenagers do not think we are particularly violent. In other words, when the responses of the students in each nation are considered, clear differences emerge. No two nations display exactly the same pattern of responses.

Overall, however, the beliefs about Americans that seem most troubling to the teenagers are that Americans are violent and materialistic and like to dominate others and that American women are sexually immoral.

## Those Violent Americans

Since the earliest days of the motion picture, Americans have been depicted as a violent people. Whether in scenes of cowboys in early movies, *Gunfight at the O.K. Corral,* Edward G. Robinson's gangster films in which cops battled bad guys with submachine guns, *Bonnie and Clyde,* or *Rambo,* Americans have been shown as trigger-happy gunslingers. Reports in the news in present times about guns fired in schools, serial snipers, and shootings in post offices and other workplaces have not softened that image. Portrayals of violence in modern films and TV programs reinforce that image.

In all but three of the countries studied, the youths believed that Americans are indeed a violent people. What they see on TV and in the movies provides clear evidence of that , and what they read or view in the news provides further confirmation. This conviction was most strongly held in two Muslim countries, Bahrain and Saudi Arabia, in which few youths have direct contact with Americans. Teenagers in Argentina, Taiwan, and Italy, where contact with Americans is more common, did not share the violent image of Americans.

### Americans as Criminals

What would the news media do if there were no crimes to report, or what if the entertainment media could no longer use criminal plots in films or TV dramas? There is little doubt that such fare adds interest and excitement to what is read or viewed. A large proportion of the "news hole" in contemporary newspapers, and of time in news broadcasts, is devoted to murders, rapes, robberies, sexual assaults, white collar crime, and the trials of apprehended perpetrators. Indeed, the content of our TV and movie entertainment makes heavy use of this theme. Little wonder that youths who develop their understandings of people who live in the United States from media depictions come to believe that many Americans engage in criminal behavior.

With the exception of Italy and Argentina, the youths studied believed to one degree or another that many Americans engage in criminal activities. There is little doubt that they have acquired this belief, at least in part, from the media. The perception that Americans are criminally inclined appears to be especially strong in the Muslim and Asian countries studied.

### Americans Like to Dominate Others

It may be inevitable that people in other countries believe that Americans enjoy dominating other people. This belief may be less a product of entertainment media exposure than of American activities around the world reported in the news. In recent decades, as the dramas of Kuwait, Afghanistan, and Iraq have unfolded, people around the world have witnessed news reports of the American military's dominating by force in those places, just as it had earlier in both world wars. The United States has clearly played a dominating role through the use of force in its affairs over a long period of time. Even when the cause was viewed as just, the legacy of these events is mistrust, which probably contributes to the overall negative attitudes that have developed.

### Americans as Disrespectful of Other People

In recent times, the phrase "ugly Americans" became popular in some countries as a way of describing how tourists and others from the United States appeared to local people. It was a clear sign that when Americans visited other countries, they had little respect for local customs of dress and etiquette and other cultural expectations. While this phrase and its meaning may have been generated by contact between gauche American visitors and local populations, it appears to have been reinforced in media entertainment. In a number of the nations studied, for example, the clothing worn by American actors, as they play their roles in dramas and motion pictures, would not be tolerated by local populations. Women on the screen who appear in revealing outfits would be in serious violation of many local dress codes. These and other personal behaviors, as depicted on TV or in movies, may seem like evidence to others that those "ugly Americans" have not changed.

### Americans as Overly Materialistic

It is perfectly obvious to audiences around the world that the lifestyles of Americans shown on screens are characterized by abundance. As seen in movies and on TV, many of the people in the United States have beautiful kitchens; they live in nice, single-family houses or spacious apartments; they drive modern cars, eat in expensive restaurants, and stay in elegant hotels. In less prosperous countries, or in nations where many people consider themselves to have strong spiritual values, such materialistic displays may be seen as morally objectionable and offensive.

The belief that Americans are very materialistic is especially strong in a number of countries. Again, youths in the Muslim and Asian countries appear to subscribe to this conviction, with some exceptions such as China and Pakistan. The teenagers in Pakistan, which is one of the poorest of the countries studied, did not view materialism in such a negative way.

## American Women as Sexually Immoral

Given the content of the majority of motion pictures that depict relationships between males and females in the United States, it is not difficult to understand that the youths studied often see American women as sexually immoral (Argentina and Italy being exceptions). Figure 4 shows that the youths in almost all the countries studied held this view. In fact, a majority of the students "strongly agreed" with the statement "Many American women are sexually immoral." Overall, this statement appeared to provoke more intense responses than any other statement in the questionnaire.

Women are often shown in Western movies and TV dramas engaging in sexual conduct that is regarded as flagrantly promiscuous in some of the countries studied. In many restrictive societies such conduct includes kissing or even holding hands in public settings. It also applies to dress. Consider the dress norms in many Muslim countries, where women who venture outside their homes wear garments that cover them more completely. In contrast, in American movies and TV dramas, viewers see women dressed in revealing clothing that would cause outrage if worn in

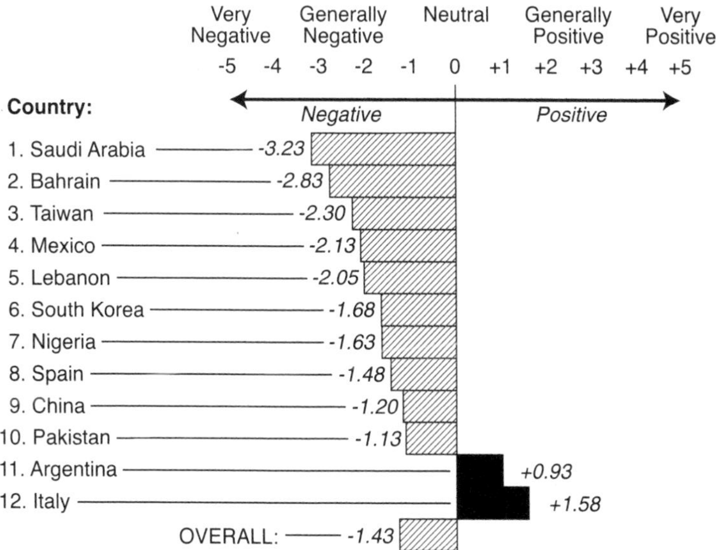


**Figure 4.** Overall attitudes toward Americans by scale item

public in at least some of the countries studied. Such entertainment also shows men and women in bed, engaging in clear depictions of sexual acts, often with people to whom they are not married. Such behavior is viewed with alarm by many in countries with conservative norms.

Thus the data shown in Figure 4 were undoubtedly influenced in part by the norms for female conduct prevailing within the societies of the countries studied. Therefore, a reasonable inference is that both media depictions and local codes of conduct influenced the youths' interpretations of the sexual morality of American women.

## DISCUSSION

Assuming that the findings of this research are valid, what are some of the possible consequences of these teenagers' beliefs about Americans? A number of implications can be identified.

### The Accusation of "Cultural Imperialism"

One consequence of the flow of information from the United States to other countries, made possible by the long-term adoption of media technologies, is the charge of "cultural imperialism."

For example, many scholars claim that, over the years, no real increase in "foreign news" has appeared in American news media. News from the United States to other countries has increased as news media have been adopted. The flow of news, in other words, is extensive *from* the United States to other nations, while a number of scholars have concluded that only a limited amount of foreign news has come back to the United States from other parts of the world.[7]

In this sense, it is quite true that American media dominate the world, providing a constant outward flow of entertainment products and daily news stories. Many critics note this lopsided movement of news and entertainment and conclude that it has important power advantages for the United States. That is, the consequence of this one-sided flow is that the United States gains political or economic hegemony over those on the receiving end.

However, to many scholars, the facts do not seem to support an imperialistic interpretation of what is taking place for several reasons. First, the term "imperialism" is an emotion-laden one implying that a powerful nation forces itself on a weaker one by some overt means in order to gain advantage and control. In the case of mass communications, it would be difficult to identify any kind of "force" that is being employed. Those nations that do import mass communication content from the United States do so voluntarily. Many critics also maintain that this one-sided flow is brought about by a conspiracy on the part of the American government and powerful economic organizations in the United States that either export the communication content or benefit from its use in other countries. However, it is clear that in the countries studied, youthful audiences greatly enjoy and demand what they receive. If those sources of news and amusement were suddenly stopped by their local governments, considerable protest would be likely. Again, what flows from the multinational producers to those in importing countries is invited and willingly tolerated, not forced on those who receive it.

On the other hand, scholars and critics also argue strongly that such media products are too compelling for those in other nations to resist. In time, attending to media portrayals can displace local values and culture. Thus, the exportation of American media products into other nations can become a source of significant resentment in those countries over time.[8]

## Influences on Foreign Affairs

There are several areas of concern about future foreign affairs that appear to be important in understanding the implications and consequences for Americans of what has been discussed in this chapter. That is, if it is indeed true that many of the young people studied of the next generation in the societies included in the study have decidedly negative feelings about ordinary Americans, what does that fact imply for the future?

One possible area of concern is the political consequences. A major issue here is the possibility of future acts of terrorism, either in the United States or in other areas of the world where Americans are located. As recent events have shown, such acts are almost universally committed by young individuals who see the people of the United States as their enemy. Another area of concern is potential business and economic consequences. The problem here is that if those of each new generation in countries such as those studied hold negative views of Americans, that could lead to a reduction in their willingness to import goods and services from the

United States. If that happens, the current favorable economic position in the world marketplace that the United States enjoys could be eroded.

*The Role of Youth in Acts of Terrorism*

It seems inevitable that negative incidents will occur in countries where there is an American presence, particularly a military presence. Military interventions, violations of cultural norms, or crimes and accidents can arouse strong negative feelings about the people of the United States and its government. However, such incidents alone do not usually provide necessary and sufficient conditions to provoke drastic terrorist actions.

What this means is that, given recent history, it appears that other conditions must be present to provoke bombings and similar destructive acts. One possibility is that a religious factor may serve as a necessary condition in some cases. In recent years, religion does appear to be a *potential* (necessary but not sufficient) condition that can lead to acts of terrorism by some militant groups.

The question is, then, what is needed as a motivating factor to enable those committed to terrorism to arouse young people to engage overtly in hostile acts even when such acts may require their own suicide? One factor appears to be the following: *There must already be in place a long-term foundation of shared negative beliefs and attitudes toward the people of the United States upon which the feelings generated by the specific incident can be based.*

In other words, as illustrated by the findings presented in this chapter, the factor of commonly shared negative evaluations of Americans may play a role in providing a third necessary (but not sufficient) condition that can increase the likelihood that at least some young people may be recruited to engage in overt acts of terrorism against the people of the United States.[9] Obviously, many factors are responsible for such an act, but it may be that a foundation of shared negative beliefs is one step in that long causal chain that may eventually result in a terrorist act. If this is the case, Americans will continue to experience anxiety and distress as they cope with alerts and news reports of possible and actual terrorist acts.[10]

*Negative Attitudes Can Influence the Marketplace*

Even though it has only 4.7 percent of the world's population, the economy of the United States has no current equal.[11] If the GDPs of all nations on earth are combined, a third of everything that is produced and brought to market on the planet comes from the United States. The official policies of the United States are designed to increase world trade and open markets for its products on the grounds that this is a sound strategy for bolstering the American economy. Traditionally the American stock and bond markets provide investment opportunities for people in many countries. Moreover, the United States is far ahead of any other country in the issuance of patents, as well as in expenditures on research and development. For that reason alone, in spite of temporary business cycles, it is likely to maintain economic world leadership in the future.

However, what would happen if a substantial part of the populations in the world decides that it no longer wants American investments or exported goods and services? The findings in the present chapter are troublesome in this respect. They suggest that a less favorable state of economic affairs for the United States could be a possible result in the future. Specifically, that segment of the world's population who are now teenagers will soon be the parents and heads of families who manage businesses and become influential in their governments. As additional generations also are added, the resulting populations will make economic decisions at all levels. Thus, if their media-shaped negative assessments of Americans continue into their adult years, there may be economic issues at stake.

Generally, then, even though people worldwide eagerly acquire our latest media products, our manufactured goods, and many other things that Americans produce, that can change. As globalization continues in the production of consumer goods, financial services, and other aspects of economies, there will be increasing competition on all fronts. If there continue to be widespread negative attitudes and beliefs about Americans, generated in part by exposure to mass communications, these factors together could result in unwanted economic consequences.

## Advice to Policy Makers, Practitioners, and Educators

According to Richard Moose, the negative image of Americans and of the United States is the most important long-term problem we face. Recent efforts have been made through public diplomacy to improve our image abroad. These include the efforts of Karen Hughes, the U.S. under secretary for public diplomacy, and those of Condoleezza Rice, U.S. secretary of state, and others.[12] Unfortunately, these efforts have not all been well received. However, there is a growing awareness of the importance of public diplomacy and recognition that public diplomacy efforts must be improved and sustained. Mr. Moose emphasizes that there isn't much we can do to change our image easily or quickly. However, he stresses, we can help our own people to understand the outside world better so that we don't do so many things that trouble others and cause many of our current problems.[13]

It is evident that movie makers and producers must think more about the long-term consequences of consistently negative portrayals of Americans in entertainment products viewed abroad. The response of industry executives that "we are just giving them what they want" is not adequate. Industry executives must also shoulder their share of ethical responsibilities and find creative ways to present more balanced portrayals over time.

Educators have responsibilities, too. Scholars and social scientists in mass communication and related fields must turn more attention to studying the effects of media content produced in this nation and consumed by people in other nations. The study of intercultural communication is becoming increasingly important to the affairs of this nation. Educators who teach children in grades K–12 (or the equivalent) the world over must help their young students to analyze and interpret media content, and to understand the degree to which it accurately—or inaccurately—reflects reality.

The negative image of Americans and of the United States is a serious long-term problem that this nation must face. If this problem is not resolved, it may have serious consequences for the United States for years to come.

NOTES

1. Iris Noble, *Nellie Bly, First Woman Reporter* (New York: Messner, 1956).

2. Joseph Turow, *Media Today: An Introduction to Mass Communication* (Boston: Houghton Mifflin, 1999), 247–48.

3. Katharine Q. Seelye, "Newsweek Apologizes for Report of Koran Insult," *New York Times,* May 16, 2005.

4. Melvin L. DeFleur and Margaret H. DeFleur, *Learning to Hate Americans: How U.S. Media Shape Negative Attitudes among Teenagers in Twelve Countries* (Seattle: Marquette Books, 2003).

5. Several students who came to the United States from conservative nations to pursue degrees at American universities have indicated this to the author. At home, they were closely governed in what they wore, how they spoke to others, with whom they could socialize, etc. These behaviors were controlled by their families and by local norms. Some reported that when they viewed American family situation comedies on TV in their home countries, they were secretly envious of the teenagers portrayed when they talked back to or disobeyed their parents or dressed in ways not acceptable in their own society.

6. The concept of "incidental lessons," whereby media audiences unwittingly learn about a topic while being entertained by media content in which the lessons were not deliberately placed by the producers, was first described by Wilbur Schramm, Jack Lyle, and Edwin Parker, *Television in the Lives of Our Children* (Stanford, CA: Stanford University Press, 1961), 75.

7. A study of nearly forty thousand U.S. newspaper articles dating from 1927 to 1997 showed that about 10 percent reported foreign events. Cleo J. Allen, "Foreign News Coverage in Selected U.S. Newspapers, 1927–1997" (Ph.D. diss., Louisiana State University, 2005), 4–15. Whether there was a trend of less foreign news was not clear. At times that was the case; at other times it was not.

8. For a discussion of cultural imperialism, see: Arthur Asa Berger, *Cultural Criticism: A Primer of Key Concepts* (Thousand Oaks, CA: Sage Publications, 1995), 61–64. See also Michael G. Elasmer and John E. Hunter, "The Impact of Foreign TV on a Domestic Audience: A Meta-analysis," *Communication Yearbook* 20 (1997): 47–69.

9. This set of conditions is not exclusive to the American situation. In May of 2003, officials in Saudi Arabia uncovered a plot to bomb the holy city of Mecca. The terrorist groups planning the incident were able to recruit individuals as young as fifteen. Three were seventeen and one eighteen. More than half the Saudi population is under age eighteen. See Faiiza Salah Ambah, "Suspects in Saudi Terror Plot Reportedly Include Teens," *Boston Globe*, June 23, 2003.

10. Jonathan Saltzman, "Far from the Front, Cases of Anxiety Rise," *Boston Globe*, April 6, 2003.

11. "The World in Figures," in *The World in 2003*, special issue of *The Economist*, November 2002, 81–87.

12. Chris Brummitt, "U.S. Envoy 'Misspeaks' during Talk," *Baton Rouge Advocate*, October 22, 2005.

13. Concluding comments by Richard Moose at the John Breaux Symposium at the Manship School of Mass Communication at Louisiana State University, November 12, 2005.

# 6

# INSTANT CONNECTION

## *Foreign News Comes In from the Cold*

JOHN YEMMA

Foreign correspondents have always been the princes of the profession. Independent and resourceful, they roam the world, pursuing conflicts and famines, coups and earthquakes. Much of what they witness and write about is profoundly serious: terror, ethnic conflict, unimaginable poverty. They also file their share of "DBIs," dull-but-important pieces, about, for example, a currency crisis in Argentina or a cabinet shakeup in Bucharest. And at one time or another, almost every foreign correspondent ends up covering a war. No one emerges from such an experience unchanged. War forces a journalist to witness suffering and death on both an intimate and vast scale. War puts a foreign correspondent into direct danger, calls into question rules about neutrality, forces improvisation, addicts with adrenaline—and, truth be told, punches a journalist's career ticket.

A foreign correspondent's life can be, by turns, vivid, tragic, comic, and terrifying. It is rarely boring. The field has its share of celebrities (Winston Churchill, Ernest Hemingway, George Orwell, Martha Gellhorn, Ernie Pyle) and controversies (Walter Duranty, Jack Kelley, Judith Miller). If there was a foreign correspondent from Central Casting, it has to have been Richard Harding Davis, the dashing scribe who chronicled the Rough Riders, covered both sides of the Boer War, and was equally at home in the wilds of the Balkans, the jungles of the Congo, and the better restaurants of Paris. A perfect day for Davis, one wag wrote, would consist of "a morning's danger, taken as a matter of course; in the afternoon a little chivalry, equally a matter of course to a well-bred man; then a motor dash from hardship to some great city, a bath, a perfect dinner nobly planned. Shrapnel, chivalry, sauce mousseline, and so to work the next morning. . . ."[1]

Even today, a stint as a foreign correspondent is likely to be the most colorful assignment of a journalist's life. Little wonder that the writing of foreign correspondents is expected to be colorful as well. He or she mines a vein of material largely unknown to readers back home. City Hall, Downtown, Capitol Hill—American news consumers are familiar enough with domestic characters and issues. But the pungent souks of Damascus, the prayer wheels of Lhasa, the sprawling favelas of Rio—these are part of the great other, the world "out there" that any true foreign correspondent longs to see and thrills to describe.

Free from the mundane aspects of journalism, far from meddling bosses, a foreign correspondent is as independent as a journalist can be. In large measure, that freedom has evolved as the by-product of necessity. For much of the twentieth century, the tether to the home office was tenuous and communication was unreliable. A foreign correspondent had to operate autonomously. Until the mid-1980s, for instance, the main connection with the foreign desk was via cable, telex, or an "onpass" arrangement with a wire service, which would forward an article to a client or subscriber newspaper.[2] The quality of overseas phone calls was spotty until the 1980s, and the cost of such calls made voice communication infrequent, especially for journalists working for smaller newspapers.

As a result, a foreign editor would attempt to direct a foreign correspondent through terse cablese.[3] So sketchy was this form of communication that the correspondent would mostly figure out the best story to write on his own, send a heads-up message, gather material (a trip to the front line, a few quick interviews), pound out eight hundred to a thousand words on a manual typewriter, and hand the piece to a telex operator to transmit. When published, the news would be at least twenty-four hours old. A foreign correspondent had to be a self-starter and a sparkling writer. Little wonder that a successful foreign correspondent often developed an air of assurance that bordered on arrogance. Editors indulged—and readers expected—a foreign correspondent's authoritative and unchallenged accounts of faraway people and events.

All that is now changing, due largely to technological innovation. The era of princely autonomy and unchallenged authority has been eclipsed by a tighter electronic tether, more widespread scrutiny and critique of content, and an almost instantaneous feedback loop that makes what was once exotic and far away seem up close and familiar to readers.

## SOURCES AND METHODS

Old foreign hands traditionally saw their skill set as comprising not only an array of special techniques for getting the story—from working with a trusted local "fixer" to fancy-footing a government "minder," from joking one's way past a checkpoint to spreading a little baksheesh—but also for getting the story home. Without a quick route back to a serviceable telex or phone, a day as an eyewitness to history is a day lost. Even the most amazing story is inconsequential if it cannot be told. Playwright Tom Stoppard nicely captured the importance of getting the story home in his 1978 play *Night and Day,* in which the journalists who take refuge in British businessman Geoffrey Carson's African villa are thrilled to discover that he has a private telex.[4] Now they could file! In the real world, Serge Schmemann, editorial page editor of the *International Herald Tribune* and longtime foreign correspondent for AP and the *New York Times,* recalls landing in Zaire in the 1970s and as the first order of business bribing a telex operator at the post office and another at his hotel to ensure timely attention to his forthcoming dispatches. Only then would he strike out for the front line.[5]

Getting the story home was such a commonplace concern until recently that foreign correspondents often joked that when all else failed they would have to send by "cleft stick"—a reference to the useless advice dispensed to neophyte foreign correspondent William Boot by the eccentric publisher in Evelyn Waugh's comic masterpiece *Scoop:*

> "There are two valuable rules for a special correspondent—Travel Light and Be Prepared. Have nothing which in a case of emergency you cannot carry in your own hands. But remember that the unexpected always happens. Little things we take for granted at home . . . like a coil of rope or a sheet of tin may save your life in the wilds. I should take some cleft sticks with you. I remember Hitchcock—Sir Jocelyn Hitchcock, a man who used to work for me once; smart enough fellow in his way, but limited, very little historical backing—I remember him saying that in Africa he always sent his dispatches in a cleft stick. It struck me as a very useful tip. Take plenty."[6]

If Hildy Johnson and the other hypercompetitive rogues of *The Front Page* captured the exaggerated truth of big-city journalism in the first half of the twentieth century, *Scoop* has long served as the comic touchstone for foreign correspondents. Bill Deedes, the famed British journalist whom Waugh used as his model for the young Boot, recalled that during the Abyssinian conflict of 1935–36, which both he and Waugh covered, Waugh "made light of our endeavors to raise stories for our newspapers in a capital where there were no stories to be had—*unless you made them up*" (emphasis added).[7] If a modern-day Deedes were to toss off such an admission of journalistic fraud, he would doubtless provoke a major credibility scandal in journalism. But the truth is that for many decades, this sort of corner-cutting was, if not the norm, certainly not unfamiliar to foreign correspondents.

*Scoop* also plays on the minimal relations between correspondent and home office that is—or, at any rate, was—a commonplace of the occupation. Boot, the hero, is a gardening writer who through a mix-up is sent to cover a war in the fictional country of Ishmaelia. He has no special qualifications for the assignment, but after a series of misadventures and upbraidings[8] from his foreign editor, he triumphs, in part because the story practically writes itself. As one of his clueless yet highly informed telegrams to his home office in London reads: "NOTHING MUCH HAS HAPPENED EXCEPT TO THE PRESIDENT WHO HAS BEEN IMPRISONED IN HIS OWN PALACE . . . LOVELY SPRING WEATHER BUBONIC PLAGUE RAGING."[9]

Boot and his colleagues operate out of the Hotel Liberty, a war-zone establishment that caters to their needs and is eager for their money. Such places have long served as forward operating bases for foreign correspondents.[10] Among the more famous names on this roster: the Holiday Inn in Sarajevo, the Continental in Saigon, and the Commodore in Beirut.[11] A well-run hotel traditionally was the best guarantee a foreign correspondent had that he would be able to write *and file* successfully. Throughout the Lebanese civil war in the late 1970s and '80s, Fuad Salah, the Commodore's resourceful proprietor, kept his electric generators running, had a brace of AK-47s stashed under the front counter to ward off marauders, and always ensured that the telex machines worked. The Commodore bar served as both watering hole and information exchange. The taxi drivers who loitered outside were trustworthy. And for good measure, there was a parrot in the lobby that did a dead-on imitation of an incoming artillery round.

If being a foreign correspondent is exhilarating and intense, it has a serious personal side as well. It can, for one thing, be achingly lonely. Bad enough that communication with the home office is minimal; meaningful contact with stateside family and friends can be virtually nonexistent. Whether in Fallujah or central Asia, Yemen or the Congo, a correspondent can easily emerge from an assignment with knowledge and insights that he realizes will be lost on anyone but fellow correspondents. In that sense, one's isolation increases with each new experience. Far from home, a correspondent can fall prey to alcoholism, infidelity, and burnout.[12] Robert Kaplan, the erstwhile foreign correspondent and author, has noted:

> You now know something vital about the world that no one else does, but it doesn't help you in your daily life. Your experience has only made you lonelier. Foreign correspondents are particularly oppressed by this sensation. For them, places like Saigon, Beirut, Sarajevo, Kabul, and lately Baghdad are not the places everyone else thinks they know through the headlines and history books, but different, far richer realities. As with Joseph Conrad's Lord Jim, it is the very complexity of the truths that they know firsthand that cuts them off from their families and the rest of the world.[13]

There can be professional consequences to independence and isolation as well. Away from bosses and colleagues, a jaded hack[14] can pick up bad journalistic habits. Among these are "embroidering," i.e., making a real event more colorful or dramatic than it was; lifting quotes from other newspapers, especially local ones, without attribution; and "piping," or fabricating characters, events, or quotes. Foreign reporting can easily fall prey to such fraud, since foreign sources until recently seldom saw their quotes in print and editors and readers back home usually had little personal knowledge of events in far-off lands.

Which leads to the gist of my argument: rapid changes in technology are ending the era of the freewheeling, independent foreign correspondent, and while there are costs and benefits associated with this change, it seems clear to me that benefits predominate. One might expect, for instance, colorful writing to suffer because a correspondent is now becoming much more connected to the home office. I believe the opposite is true: correspondents are being freed from the logistical problems of filing over shaky local infrastructure and can travel farther, faster, and for a longer period of time into a foreign countryside and culture. They need not gather at one war-zone hotel, since their communication technology is increasingly portable. Technology is also removing the isolation that foreign correspondents are susceptible to and opening a field once seen as an elite calling to all manner of journalists and would-be journalists. Most importantly, technology is enabling vital journalistic safeguards involving truth and accuracy to be extended to foreign reporting.

## FASTER, BETTER, CHEAPER

Depending on the culture of the news organization, a foreign correspondent's prose until recently was either untouched by the editors back home or edited with little or no consultation. A reporter wrote, filed, and moved on. Beginning in the early 1970s, when computers were put into service in foreign and domestic bureaus of news services and newspapers, information technology underwent a steady evolution. Initially, these changes added efficiency to writing and editing, but soon they yielded more efficient ways to communicate between correspondent and editor as well. The impact of this technological quickening has been profound, especially in the way it has freed foreign correspondents from having to find a decent telex or phone line.[15] There is almost no time today when a foreign correspondent is out of touch, and rarely do reporters have to make filing a major consideration in their workdays.

Ethan Bronner, deputy foreign editor of the *New York Times,* began his foreign reporting career with Reuters at its Fleet Street headquarters in 1980 at a time when stories were written on manual typewriters with carbon copies. In subsequent assignments in Madrid, Brussels, and Jerusalem over the next five years, he recalls, "There was never any back and forth over editing."[16] Over the years, as one form of technology displaced another—from clattering teletypes to the fifteen-pound Teleram Porta-Bubble/81 computer to satellite phones, e-mail, and global Internet—Bronner and other journalists have seen the relationship between desk and correspondent grow closer. By the time he returned to Jerusalem in

1993, he would talk over his assignment with an editor almost every day by phone and e-mail.[17]

The upside of such technological advances is self-evident, Bronner says: "Lots of people benefit from talking through a story in advance."[18] Most current foreign correspondents and foreign editors agree. Typical is James Smith, whose career includes time with the Associated Press and the *Los Angeles Times* in posts ranging from Tokyo to South Africa to Mexico City. Smith started in the era of the teletype, graduated to the Radio Shack TRS-80 (the famed "Trash 80"), and has lived through many generations of ever-more-portable, ever-more-reliable computers throughout his career. As foreign editor of the *Boston Globe,* he pioneered the use of Skype, the instantaneous voice over Internet protocol (VoIP) service, to keep in touch with his bureaus. There's no doubt, says Smith, that accuracy is much greater, that research and sourcing are better, and that stories are more completely thought through because of enhanced communications technology.[19]

Marjorie Miller, foreign editor of the *Los Angeles Times,* recalls her days of filing on a TRS-100 (successor to the workhorse 80 and the first flip-screen laptop) from El Salvador as a time when the logistics of filing consumed large amounts of her day. Miller says she could devote only two or three hours to reporting and the rest of the day to trying to find a good phone line so she could attach alligator clips or an acoustic coupler to the handset to transmit her story. "Filing was a big deal," she recalls.[20]

As Miller's experience underscores, far from reducing the opportunities to absorb local color, new technology, in theory at least, should be increasing those opportunities. Miller, who started her career as a local hire in Mexico City in the early 1980s, joined the *LA Times* in 1983 and, besides El Salvador, was stationed in Mexico, Israel, Germany, and Britain. The last war she covered was Kosovo in 1999, and by then she could set up a satellite phone in a farmer's field to file her story. It was still bulky equipment (eight to ten pounds), but it meant that she did not have to rely on hard-wired phones and telexes in a major city. Today, she says, reporters at the *Times* are in constant contact.

While technology has definitely improved the speed of reporting, the new ease of access that the home office has to a foreign reporter means there are also more demands on the time of reporters. In hot spots like Iraq, reporters are expected not only to file daily pieces but also, increasingly, to contribute to their paper's Web site during the day. They may also do interviews with National Public Radio and other broadcast outlets to increase their paper's exposure. And while the Internet helps reporters know what their competition is doing, it sometimes causes them to worry too much about the competition, says Miller. The endlessly evolving world of information on the Net means that some reporters are not out on the street as much as they would have been in the past.

For his part, Smith notes that a correspondent no longer can "claim to be out of touch"—which is not necessarily all to the good. "The contemplative quality of absorbing a place for days at a time without worrying about filing" may have been lost, he says. He also acknowledges the "multiple distractions" brought on by constant access to the Internet from anywhere in the world.[21] Bronner, who works with by far the largest staff of veteran foreign correspondents in journalism today, adds that constant connectivity has caused some of the more veteran foreign correspondents at the *New York Times* to lament the loss of distance and independence.

But then, in almost every occupation the cliché about the old days is that they were more colorful, more arduous, and more exciting than today. It is true that before e-mail and satellite phones, foreign correspondents did excellent work with little hand-holding from their editors. The best of them used their freedom to range widely, filing original stories with local flavor that informed, moved, and delighted readers back home. As John Hohenberg wrote of Ernie Pyle's World War II coverage, "[He] looked for the story of the individuals swallowed up on the massive battlefront, and he took his time about gathering the material for his work. He was respectful of deadlines, but never enslaved by them."[22]

Technology is neutral in this argument. In 1914, Richard Harding Davis scorned the new era that was dawning in which "the correspondent moved with a cable from the home office attached to his spinal column, jerking him this way and that."[23] It was all so free before that, he believed. Half a century later, the clattering telex undoubtedly dismayed veterans used to the elegant simplicity of cable—and so on through e-mail, faxes, satellite phones, and VoIP.

## BETTER FOR JOURNALISM

Connectivity is an extremely important tool in safeguarding the credibility of foreign reporting. Until recently, sources in foreign stories rarely saw their words in print. How could they know whether they were quoted accurately? And skeptical readers had little comparative information for challenging fact or bias in a dispatch from far away. Editors had to trust their correspondents to be ethical. For the most part, the trust was justified. Nevertheless, when a reporter made an error—or worse, made up a quote or conflated events—there was no audit mechanism to check the facts, other than a seasoned foreign editor or a rare whistleblower.[24] Thus, journalistic abuses by foreign correspondents could go undiscovered much longer than could those committed by domestic correspondents. Jayson Blair of the *New York Times* lied in print over a four-year period. By contrast, a *USA Today* investigation of the fabrications of star foreign correspondent Jack Kelley found that he had made up people and incidents for more than a decade before being caught.[25]

Chastened by such scandals, editors are more vigilant these days, and foreign correspondents are more careful. Technology is bringing to bear better research and sourcing, valuable brainstorming between editor and reporter, more scrutiny of news reports by knowledgeable readers, and lower barriers to entry into the world of foreign correspondents.

Let's take these point by point:

*1. Stronger sourcing.* Except for time-zone differences, a journalist in Baghdad or Beijing is now in almost exactly the same information loop as a coworker on the city desk or Capitol Hill. Even in war zones or in remote areas like the eastern Congo, correspondents stay in touch via international cell phones, satellite phones, and e-mail. The Internet and various subscription databases like LexisNexis and Factiva also give a traveling reporter the same global archive and library that anyone at home can enjoy.

Consequently, foreign stories today are scarcely limited to good datelines and one or two local quotes. Reporters are expected to use their phones to enrich their reporting.[26] Necessity plays a part in the interconnectivity, too. At most news organizations, reference libraries, which once backstopped traveling reporters, are a shadow of their former selves, and the ranks of foreign desk editors has been thinned as well. The Internet must fill the gap.

2. *Improved communication.* There is no doubt that the tether is tightening in ways that alter the news gathering process. E-mail and instant phoning are the modern manifestation of Richard Harding Davis's complaint about the cable from the home office being attached to a reporter's spine, although each reporter and editor is different in this regard.[27]

If, on balance, new technologies drastically decrease the autonomy of foreign correspondents, they also ensure that editing is more cooperative, transparent, and iterative. A correspondent can no longer complain that "Boston" or "Chicago" or "LA" took liberties with his or her prose when the correspondent was out of touch. Hard questions can be asked of a correspondent on the other side of the world when they need to be asked: before publication. Because of this, foreign correspondents have become stronger at sourcing—pushing for named sources rather than relying on "western diplomats" and other dubious old standbys—and are less likely to take liberties with descriptions and quotes than in the past.[28]

3. *More scrutiny. Los Angeles Times* foreign editor Marjorie Miller agrees that "all of the contact probably makes it harder to invent" people and situations in the way that Bill Deedes laughingly described. She is not sure, however, that that technology is necessarily the solution to the problem of journalism fraud, since the Internet can also make it easier to crib information from other sources.[29] But when a subject or source in Nairobi or Tblisi can go to a Web site and read a foreign correspondent's story the next day, the piece gets the same sniff test that domestic stories get. This should improve accuracy in foreign reports. It might also make foreign correspondents more cautious, knowing that their subjects may well read articles written about them and take issue with them. In any case, it puts foreign reporting on the same playing field with domestic reporting.[30]

In addition, reporters in the same faraway locale—whether cooperating or competing—can read each other's published dispatches, ensuring a further level of scrutiny. And because the stories live virtually forever in cyberspace, rather than moldering in a newspaper morgue, a foreign correspondent, like his domestic counterpart, becomes subject to the scrutiny of press critics, bloggers, and media writers. In August 2006, Charles Johnson, who blogs under the name "Little Green Footballs," blew the whistle on a doctored Reuters photo of a Beirut neighborhood after an Israeli air strike. Reuters subsequently retracted the photo and suspended the photographer.[31]

Life on the road is becoming less autonomous, but after Jack Kelley, Jayson Blair, Stephen Glass, and too many other journalistic scandals, this is an inevitable and necessary development. The Kelley case is instructive in this regard.[32] Bill Kovach, who along with Bill Hilliard and John Seigenthaler led the outside investigation of Kelley, says he believes new technology is helping clamp down on journalistic miscreants in several important ways. He acknowledges that, at first blush, Kelley's "piping" techniques were in fact enhanced by technology (as indeed were Jayson Blair's): they would download stories from other publications and use them as "fodder to cheat." But, says Kovach, "once we were on the trail of it," the Internet "made it possible to nail him in ways we couldn't have before. We used the Internet, e-mailed overseas, read stuff on line." Kovach says that when coupled with a culture of accountability and accuracy in the newsroom, "technology can let you spot problems faster and catch them quicker."[33]

*4. Lower barriers.* An old foreign hand knew the ropes, knew who to talk to in every country, how to craft the story, and how to file. Technology helps demystify the job. An Atlanta-based medical reporter sent to Africa to investigate an Ebola outbreak doesn't need to learn to punch a telex tape, bribe an operator, or race back to Reuters before the bureau shuts down for the day. This ad hoc foreign correspondent could e-mail officials at the Centers for Disease Control and the World Health Organization, file via satellite phone, make travel arrangements via the Internet, even stay up on office gossip via e-mail. Such technology can't hurt the veteran foreign correspondent and definitely facilitates the work of the nonveteran brought in as a specialist.

After 9/11 and the invasion of Iraq, a new paradigm appeared at American newspapers. Reporters throughout the paper were dispatched to war zones, and reporters already posted in foreign bureaus were deployed to the hot spots of south Asia.[34] To this group can be added hundreds of thousands—even millions—of amateurs and once-and-future journalists who post reports and digital photos on the Internet from war zones, natural disasters, and other events worldwide.

In the aftermath of the December 26, 2004, Indian Ocean tsunami, bloggers provided some of the first and most vivid accounts of the devastation.[35] They also were important early reporters after the July 7, 2005, terror bombings on the London underground and from the Gulf Coast after Hurricane Katrina. Rarely are bloggers disciplined enough to pursue an objective style of writing over the long run. But with no overhead—no corporate headquarters, editing staff, or printing presses—blogging has in effect made everyone a potential news gatherer. And as print journalism migrates increasingly onto the Web, the line between "professional" journalists and bloggers is becoming increasingly blurred.[36]

In an essay on challenges faced by the mainstream media, Richard A. Posner described the "blogosphere" as "not 12 million separate enterprises, but one enterprise with 12 million reporters, feature writers, and editorialists, yet with almost no costs."[37] A foreign correspondent, therefore, can venture to fewer and fewer places where he is the single, unchallenged witness to an event. Indeed, as David D. Perlmutter and Kaye Trammell point out elsewhere in this book, while "foreign blogs will [not] replace foreign correspondents, . . . they will enrich the information the world receives about important events. Specifically, these personalized posts dispatched from the front lines provide an insight and tell a story that an American deployed to the event cannot."[38]

5. *Global exposure.* The Internet exposes foreign reporters, officials, intellectuals, business people, and citizens of other countries to the journalism model pioneered by American reporters. They may reject it on philosophical, ideological, or technical grounds, as they have over the decades, but they can't say they never see it. Hard news based on primary sources is much more difficult to brush aside than opinion or subjective journalism. Fact-driven journalism that is transparently sourced and includes verbatim quotes is far and away the primary source of news that is picked up by the wire services and that echoes through broadcast media such as CNN and the multifarious channels of the World Wide Web, and that informs the actions of governments, aid agencies, and citizens around the world.[39]

Media analysts such as Stephen Hess at the Brookings Institution warn against making too much of this point. There is little evidence, they say, that the rest of the world is simply adopting the American print journalism model.[40] But even if the American model is not colonizing the world, creating local versions of the *New York Times* and *Washington Post,* the Internet and global television are amplifying the reach of the mainstream media's product. Hess, who is finishing a book on the work habits of foreign correspondents from other nations who are stationed in the United States, notes that because of time zone differences the Paris- or London-based editors of correspondents working in Washington use the Internet to read the *New York Times* five hours before their own U.S.-based correspondents do. Moreover, those ubiquitous blogs rely on the mainstream media for the bulk of their material.[41] Thus, at least for now, American media continue to set the agenda globally.

## CONCLUSION

Technology is never a panacea. It certainly is no substitute for the enterprise and diligence of a smart reporter or the experience and skepticism of a smart editor. But to the extent that technology fosters greater accuracy in foreign reporting and brings more variety of journalists onto the global stage, it can't help being a good thing.

Communication and feedback are increased when a foreign correspondent is tethered more closely to the foreign desk. Editing becomes more rigorous; the same no-cheating standards expected of a domestic reporter apply to the foreign reporter. The only local flavor that is lost in the technological revolution is the kind that journalism is well rid of: the ability of a foreign correspondent to cut corners, shave the truth, or in the worst cases fabricate people and events in a location so remote that no one is able to mount a challenge. Moreover, technology enables talented journalists, amateurs, and ordinary citizens to become ad hoc foreign correspondents, to fact-check foreign news, or to join in the global discussion about international issues. As the ranks of medium-sized newspapers able to support a foreign staff continue to fall, freelancers, bloggers, and one-time witnesses to global events will of necessity be providing much of the news that full-fledged staffers once did.

If today's world is interconnected, wireless, easy to navigate, and virtually cost-free to publish, it is no less colorful for that. There will always be a comparatively few individuals who are equipped with a passion for international issues, curiosity about fellow humans, rigorous integrity about accuracy, and fairness and skill in capturing the events of the world in words, images, and audio files. Not everyone has that skill set. Those who do will still deserve to be called foreign correspondents.

NOTES

1. Arthur Lubow, *The Reporter Who Would Be King: A Biography of Richard Harding Davis* (New York: Scribner, 1992), 306.

2. See Richard M. Harnett and Billy G. Ferguson, *Unipress: United Press International, Covering the 20th Century* (Golden, CO: Fulcrum, 2003), 204, for a good account of the transition from teletype to early computers at UPI, where I worked in the early 1970s. Although UPI and AP embraced computer technology at roughly the same time, UPI, by virtue of its underdog status, had to move rapidly because it was hoping to gain competitive advantage on its archrival. (In the end, of course, UPI failed and now it is barely a shadow of its former self.) Like those of several other contemporary journalists whom I've interviewed for this essay, my career in both domestic and foreign news has coincided with increasing technological transformation in the business.

3. When I was stationed in the Middle East for the *Christian Science Monitor* in the early 1980s, a typical service message from my foreign editor, David Anable, would read: "PROYEMMA BEIRUT EXANABLE BOSTON. TNX LATEST EXBEIRUT. COULDST FILE KURDS FOR FRIDAY DATES BEFORE PROCEEDING AMMANWARD? ALLBEST." Translation: "To: John Yemma in Beirut, From: David Anable in Boston. Thanks for your latest piece out of Beirut. Could you file your piece on the Kurdish situation in time for us to use for the Friday paper before you fly to Amman?" Even in those days there was no reason to be using such a terse construction, since telex charges were essentially like phone charges: time-based, not—like the old telegraph—word-based. But the telegraphic style persisted then as it does in some quarters today. To end an e-mail with "tnx et rgds" is to show that you are a member of the cablese fraternity.

4. Tom Stoppard, *Night and Day* (New York: Samuel French, 1978). The money quote from Stoppard's play comes from the photographer George Guthrie, who at one point in a discussion of the merits and shortcomings of journalism proclaims: "People do awful things to each other. But it's worse in places where everybody is kept in the dark. Information is light. Information, in itself, about anything, is light."

5. Serge Schmemann, interview by the author, July 26, 2005. Technological changes in recent years have been good for foreign correspondents "in every conceivable way," Schmemann says, though he has noticed that some newer reporters have never developed the resourcefulness that veterans had to have. Not long ago, he said, "a reporter called from Africa and said he couldn't file because his computer was down. I told

him to go to his hotel room, pull an envelope out of the desk drawer, take out a pen, and get to work writing. Journalism is the same whether you dictate a story or send it by telex or computer. Some reporters, however, lose the ability to function without a computer."

6. Evelyn Waugh, *Scoop* (Boston: Little Brown, 1937), 55.

7. William Francis Deedes, "Evelyn Waugh in Ethiopia: Reflections and Recollections," *Journalism Studies* 2, no. 1 (February 2001): 27–29. Deedes colorfully expands on this point throughout his book *At War with Waugh: The Real Story of Scoop* (London: Macmillan, 2004).

8. At one point Boot receives this cable from his editor: "CABLE FULLIER OFTENER PROMPTLIER STOP YOUR SERVICE BADLY BEATEN ALROUND LACKING HUMAN INTEREST COLOUR DRAMA PERSONALITY HUMOR INFORMATION ROMANCE VITALITY" (Waugh, 151). In other words, Boot was floundering in every way. Every foreign correspondent dreads such a "rocket" from the home office. Even a note that features nothing more devastating than faint praise become grist for intense interpretation to the reporter far from home.

9. Waugh, 208.

10. As longtime AP rover Mort Rosenblum put it in his classic book on the profession, *Coups and Earthquakes,* foreign correspondents "can dump a bag on any hotel-room bed in the world and feel as if they have been reared there." Mort Rosenblum, *Coups and Earthquakes: Reporting the World for America* (New York: Harper and Row, 1996), 26.

11. More recently, the Baghdad hotels that served journalists were, sequentially, the Rachid, the Palestine, the Sheraton, and the Hamra. A July 29, 2005, article by Reuters correspondent Luke Baker noted: "After the capital fell in 2003, [the Sheraton] was full to overflowing, the bar was crowded, the glass-fronted lifts zipped up and down the 19 floors, and everywhere opportunity glinted. There was the occasional mortar or rocket attack, including one that severed an elevator cable, sending the carriage crashing to the lobby. But in spite of that, it enjoyed a postwar heyday, becoming a lively hub at the heart of the country's chaos. Now it is all but dead, a symbol of how Iraq's worsening insurgency is sucking the life out of the city."

12. As anthropologist Ulf Hannerz noted in his 2004 study of foreign correspondents, "The continuous psychological stress of going on and on gathering news and getting one's writing done at a fast pace, in a setting where things keep happening, can wear a person down—especially when news work turns into a matter of personally witnessing violence and disorder." See Ulf Hannerz, *Foreign News: Exploring the World of Foreign Correspondents* (Chicago: University of Chicago Press, 2004), 99.

13. Robert D. Kaplan, "Get Me to Vukovar: The Lure of the Dangerous Road," *Columbia Journalism Review* 43 (September–October 2004): 11.

14. In other contexts, a "hack" is a derogatory term for a churn-it-out writer. Foreign correspondents—even the most literate and polished of them—call themselves

hacks in a self-mocking way. There is, of course, some aptness to the term, since a journalist is writing for money, not art, every professional day of his life.

15. Getting to a working telex or phone line was all-important in my first posting to the Middle East in the early 1980s with the *Christian Science Monitor.* During the 1982 Israeli invasion of Lebanon, for instance, three or four fellow reporters and I would hire a taxi in the morning and race to the front lines to interview soldiers and civilians. By midafternoon we'd have to double back to Beirut so that we could write and file with the hope that the telex lines were still functioning. There was simply no other way to get our reports to our editors back home. Half the day would be lost in travel and hours more lost in the effort to transmit. By 1990, when I returned for the first Gulf War as a reporter for the *Boston Globe,* I had a portable computer. That made writing easier and obviated the need to go to the lobby of the hotel to send a telex, but I still had to race back to the hotel to get a good phone line in any country I was covering (I shuttled among Jordan, Saudi Arabia, Iraq, and Egypt). Throughout the 1990s and the crises in Somalia, Rwanda, Haiti, and Bosnia, technology improved so much that, as foreign editor, I could be in daily phone contact with reporters, though even then they were usually taking the call at a stationary location. Cell phones and satellite phones arrived in the mid-'90s; shaky and unreliable at first, they nonetheless began the process of freeing reporters from their dependence on hard lines. By the time of the Kosovo campaign in 1999, and then in Afghanistan in 2001 and Iraq in 2003, reporters were able to dial directly on satellite phones from battle zones, from the wilds of northern Afghanistan or Kurdistan, from Humvees in which they were riding as embeds.

16. Ethan Bronner, interview by the author, June 20, 2005.

17. This was pre-Internet e-mail: basically, an internal e-mail system that Bronner and other correspondents accessed via the private Infonet service.

18. Bronner interview.

19. James Smith, interview by the author, June 20, 2005. Smith was foreign editor of the *Globe* until early 2007, when the *Globe,* like many other medium-sized newspapers, was forced to shut its last three foreign bureaus amid deepening financial difficulties. Even with the faster, cheaper, better technology available to foreign correspondents today, the cost of supporting reporters overseas has proven untenable for all but the largest news organizations.

20. Marjorie Miller, interview by the author, June 24, 2005.

21. The celebrated but possibly apocryphal story in the 1970s and '80s was the foreign correspondent who was talking with his foreign desk on a scratchy phone line from his Cairo hotel room and did not like what he was hearing. So he put the receiver up to his portable shortwave radio, which he jammed between stations, and shouted, "I'm losing you," before hanging up and retiring to the lobby bar for the evening.

22. John Hohenberg, *Foreign Correspondence: The Great Reporters and Their Times* (New York: Columbia University Press, 1964), 359.

23. Lubow, 289.

24. Stephen Bates, *If No News, Send Rumors: Anecdotes of American Journalism* (New York: Holt, 1989). Among the more notorious fictions in the world of foreign correspondence were the 1983 "Hitler diaries" hoax that *Newsweek* fell for and the 1981 piece on Cambodia in the *New York Times Magazine* by freelance writer Christopher Jones. Decades earlier, during the intensely partisan conflict of the Spanish Civil War, Claud Cockburn of *The Week* and Arthur Koestler of the *London News Chronicler* filed fictitious reports out of sympathy with the loyalist cause. That was a different era, however, and as bad as those transgressions were, they were not as devastating to the credibility of journalism as the relatively recent frauds committed by Jayson Blair of the *New York Times,* Jack Kelley of *USA Today,* Stephen Glass of the *New Republic,* columnists Patricia Smith and Mike Barnicle of the *Boston Globe,* and, earlier, Janet Cooke of the *Washington Post.* Before the 1950s and 1960s, accuracy and objectivity were not the central tenets of journalism that they later became.

25. Colleagues had raised warnings about both Blair and Kelley well before they were caught. Both were stars at their papers, however, so they were given the benefit of the doubt. When the whistle was blown on Blair, it was relatively easy to examine his record of domestic reporting. In the case of Kelley, however, the reports from abroad were more difficult to track down, and Kelley's attempts to cook up verification were harder to overcome.

26. A good example can be seen in the June 20, 2005, *Boston Globe.* Leading page one was a report from Pretoria by John Donnelly in which he described problems that had cropped up in getting generic AIDS drugs to Africans. Donnelly's piece featured phone interviews with the head of the Global AIDS Alliance in Washington, DC, the head of HIV/AIDS programs for the Geneva-based World Health Organization, the traveling deputy coordinator of the U.S. global AIDS program (whom Donnelly reached by phone in Mozambique), a senior executive with a South African pharmaceutical company in Cape Town, the chairman of Uganda's National Drug Authority in Kampala, and the head of Nigeria's national antiretroviral committee in Lagos. Regulators in Ethiopia and Tanzania, Donnelly reported, did not return phone messages.

Donnelly says technology "has made my job so much easier in terms of getting hold of people, while also raising the bar in terms of expectations of what stories should include. Both, obviously, are great developments." In parts of Africa where eight or nine years ago he would have been cut off from the world, he says, "I can now access people globally, because they also have cell phones, e-mail, and BlackBerries." The June 20 story, he said, was "like almost every one that I do. I used a mix of high-tech assistance, along with time-tested reporting techniques, like face-to-face reporting, and building up sources over the years." He concludes, "Everything comes together so much faster now because of these technological tools." John Donnelly, e-mail interview, June 22, 2005.

27. As Stephen Hess of the Brookings Institution noted in his study of foreign correspondents in the early 1990s, "Some correspondents complain of too much communication with their editors, but others complain that there is not enough. Sometimes . . . these reporters work for the same organization. Later back in their home office, an editor told me that variations are not based on age or experience. Some very new correspondents rarely call in, and some old timers are on the phone every day." Stephen Hess, *International News and Foreign Correspondents* (Washington, DC: Brookings Books, 1996), 66. Hess's work nicely captures the world of foreign correspondents just before the explosion in technology that occurred with the Internet, e-mail, and cell phones in the mid-1990s. He concludes his chapter on technology with a quote from the *Washington Post*'s then-managing editor Robert J. Kaiser, who lamented that because "the distance between home office and correspondents has shrunk dramatically, [it is not] as much fun overseas as it once was" (67).

28. Walter Duranty, who was stationed in Moscow for the *New York Times* in the 1930s, has come under intense criticism in recent years for his dispatches downplaying the seriousness of the Stalin-era famine in 1932–33—in effect, acting as a mouthpiece for the Soviet regime. Given the communications difficulties of the time and the extreme restrictions imposed on the foreign press by Soviet authorities, Duranty's reports went largely unchallenged by colleagues and editors. (In more recent times, Judith Miller enjoyed such prominent status within the *Times* that the sourcing of her reports about weapons of mass destruction in prewar Iraq went largely unchallenged. Technical improvements in communications would not have made much difference in her case. What the *Times* appears to have needed instead was an old-fashioned journalism-management tool: subjecting even star reporters to rigorous editing.)

29. Miller interview.

30. The August 2, 2005, murder of freelance journalist Steven Vincent in Basra, Iraq, may have been retaliation for an opinion piece he had published two days earlier in the *New York Times* in which he raised alarms about the Basra police force's carrying out political assassinations. If it is ever determined that the people he wrote about saw his piece on the Web and then decided to kill him, this would be a troubling example of the dangers that the Internet has brought to foreign reporting—dangers not unlike the ones that have long lurked for crime reporters who wrote about the mafia. After *New York Times* Iraq reporter Fakher Haider was murdered in Basra in late September 2005, *Times* foreign editor Susan Chira told the *New York Observer:* "We believe that insurgents, or whoever these people are, read web sites" (*New York Observer,* September 26, 2005, 1). The circumstances surrounding both Vincent's and Haider's murders were murky, however; there could be other explanations.

31. A good account of the fake Reuters photo controversy can be found in Randy Dotinga, "A Blogger Shines When News Media Get It Wrong," *Christian Science Monitor,* August 9, 2006. In that article, Dotinga writes: "Armed with a tip from a reader

Johnson was apparently the first person to prove that an Aug. 5 Reuters photo of Beirut after an air attack had been faked. Johnson revealed how images of buildings and smoke had been copied to other parts of the photo, apparently with the help of software. The changes enhanced the smoke above the city, making the photo more vivid. Reuters retracted the photo and 919 others by freelance Lebanese photographer Adnan Hajj, including a second doctored one."

32. *USA Today*'s official report found that Kelley's "journalistic sins were sweeping and substantial" and turned up evidence that "strongly contradicted Kelley's published accounts that he spent a night with Egyptian terrorists in 1997; met a vigilante Jewish settler named Avi Shapiro in 2001; watched a Pakistani student unfold a picture of the Sears Tower and say, 'This one is mine,' in 2001; visited a suspected terrorist crossing point on the Pakistan-Afghanistan border in 2002; interviewed the daughter of an Iraqi general in 2003; or went on a high-speed hunt for Osama bin Laden in 2003." While it is true that most of his fraud took place within the era of the Internet, there should now be enough concern among editors, readers, colleagues, and sources that both official and amateur auditing of news reports will be taking place.

33. Bill Kovach, interview by the author, June 23, 2005. Kovach also acknowledges the rise of bloggers as journalism watchdogs and thinks this is a good thing. He was contacted recently by a blogger who had noticed an elision in a quote that was used by a columnist to mock a prominent public official. In the past, such a manipulation would have gone unchallenged; now, however, bloggers are hot on columnists' trails.

34. Stephen Seplow describes this new approach: "Moscow correspondents, Rome correspondents, Johannesburg correspondents—almost all foreign correspondents have become in effect general assignment reporters for international stories. The job requirements may not have changed—ingenuity, courage, a bit of blarney, and the ability to tell a story. But the mind-set is different. Reporters need to know more about more subjects; they need to be more flexible, more agile, more prepared to lug their SAT phones and laptops to scary places they may never have thought about." See Stephen Seplow, "G.A.s for the World," *American Journalism Review* 25, no. 7 (October–November 2003): 40.

35. John Schwartz, "Blogs Provide Raw Details from Scene of Disaster," *New York Times,* December 28, 2004.

36. A thorough (if somewhat breathless) argument detailing the rising power of blogs and the "crisis" in mainstream journalism can be found on the Jan. 15, 2005, Pressthink Weblog posting of Jay Rosen, longtime media critic and professor of journalism at New York University: http://journalism.nyu.edu/pubzone/weblogs/pressthink/2005/01/15/berk_pprd.html (accessed January 2005). A more measured view is taken by Louisiana State University's David Perlmutter ("Will Blogs Go Bust?" *Editor and Publisher,* Aug. 4, 2005, http://www.editorandpublisher.com/eandp/article_brief/eandp/1/1001009362). Blogs, he writes, "are indeed a democratic wonder—the first

instance in human history where an ordinary individual can communicate her or his thoughts to the entire planet, instantly, and without editing by the elites of journalism or government. They provide a real service, as a complement and a watchdog to the mainstream sources of information and analysis. But the cause of blogging is not helped by unwarranted and blind enthusiasm about their success that ignores the threats to their authenticity and independence."

37. Richard A. Posner, "Bad News," *New York Times Book Review*, July 31, 2005.

38. See chapter 4 in this volume.

39. Serge Schmemann of the *International Herald Tribune* sees the effect of technological evolution as especially profound in broadcast journalism. The "CNN effect," he says—that is, the global influence that twenty-four-hour CNN-style news has had over the past twenty-five years—has caused TV operations worldwide to adopt the model. Al Jazeera, he points out, knew that to develop into the pan-Arab CNN it would have to adopt the western model rather than the more traditional Arab TV approach of showing leaders shaking hands in their palaces. Schmemann interview.

40. Stephen Hess, interview by the author, June 12, 2005. Hess does, however, concur with Schmemann that the basic elements of television journalism as first developed in the United States have spread globally. The role of an anchor, the reliance on action-packed footage, the look of a TV news set, even the split-screen remote interview—all American innovations—are now industry standards, seen on broadcasts as varied as those of Al Jazeera and Japan's NHK.

41. Posner. Specifically, Posner says, "The bloggers are parasitical on the conventional media. They copy the news and opinion generated by the conventional media, often at considerable expense, without picking up any of the tab."

# 7

# HAPPY LANDINGS

## *A Defense of Parachute Journalism*

EMILY ERICKSON AND JOHN MAXWELL HAMILTON

"It's not news when a plane lands safely." That is a useful newsroom aphorism to explain what is and what isn't news. But an exception to the rule occurred on October 19, 1936.

That day *New York World-Telegram* reporter H. R. Ekins beat his rivals from the *New York Times* and the *New York Journal* in an around-the-world race, all by commercial planes.[1] He set a new record, eighteen days, fourteen hours, fifty-six minutes, and two-fifths of a second. Cameras whirred to record his triumphant arrival home. He was whisked to the mayor's office. His book, *Around the World in Eighteen Days and How to Do It,* memorialized the feat.

While Ekins was having his adventure, other air landings were making news. Two days before his return, aviatrix Jean Batten, who had been commissioned by the *New York Times* to cover a flight of her own, became "the first woman to have flown solo from England to New Zealand and . . . lowered the previous best time for crossing the Tasman Sea."[2] In early October Ekins's boss, Roy Howard, and two other publishers were among the passengers on the first *China Clipper* flight from San Francisco to Manila.[3] The trip was so sensational Warner Brothers made *China Clipper,* a feature film with Humphrey Bogart in the starring role.

The newsroom frenzy over air travel seems quaint today. Travel by plane is commonplace, even boring. Instead of being glamorous, it is full of the indignities that emerge when an elite pastime becomes one for the masses. Hence the newsroom aphorism about safe landings and the news. But if routine departures and landings of international flights do not count as news today, the connection between air travel and news is scarcely over. Just the opposite—it is taking off.

With air speeds growing ever faster and tickets growing ever cheaper in real terms, the airplane has become a significant news gathering technology. It permits correspondents to reach a breaking news story quickly and efficiently virtually anywhere on the planet. This prosaic travel has a far more profound impact on foreign reporting than any of the aerial news-making high jinks of Ekins's era, although the phenomenon has received little serious scrutiny and, when it is discussed, is usually dismissed with knee-jerk disdain. Philip Seib, a thoughtful contributor to this book, asserts in his chapter that the style of journalism in which "a reporter might arrive promptly and report breathlessly . . . is intrinsically misleading." The ability to drop in on a foreign story, the critics charge, has degraded foreign news.

The common term for this type of reporting is *parachute journalism.* The basic critique runs like this: Increasingly concerned about the bottom line and shareholder profits, news organizations want to cut costs wherever they can. One of the most expensive news gathering undertakings is posting large numbers of foreign correspondents permanently abroad.[4] The cheaper solution is to rely on modern air travel and dispatching reporters from the home office to cover must-do stories abroad. Unfortunately, the critics say, parachute correspondents "end up in places they have never seen before, with no knowledge of the language, the customs or the background to the story they are covering."[5] As a result they concentrate on violence and other sensational events, not on causes and consequences. News organizations save money, but at the cost of adequately informing their readers, viewers, and listeners.

In his recent memoir, Tom Fenton, who worked abroad for the *Baltimore Sun* before a long career overseas with CBS News, summarizes the feelings of veterans about parachute journalism: "American news organizations had so depleted the ranks of hard news reporters over the years that they suddenly had to send out whatever lifestyle, fashion, and gossip types they could muster on a moment's notice."[6]

This chapter challenges that critique—or at least part of it. Without a doubt, it is essential that major metropolitan newspapers, magazines, and networks station foreign correspondents abroad for long-term assignments. A nation with the global power and the global vulnerabilities of the United States needs experienced on-the-ground journalists spotting trends and explaining them. Foreign correspondents are an early warning system for the public and policy makers. Cutting back on their numbers is as much a national security issue as reducing our naval fleet. To this extent, the critics of parachute journalism are right to fret.

But there is more to parachute journalism than the standard critique suggests. Critics portray parachute journalism one-dimensionally, when in reality the technique comes in a number of forms that enrich both the quality and quantity of overseas news coverage. Also, critics' nostalgia about the foreign correspondence of yesteryear tends to be ahistorical. Drawing on our historical studies of foreign news and on some fifty interviews with editors and reporters, we argue parachute journalism has merits deserving serious inquiry given its potential for enhancing coverage. We approach the task by looking at historical factors bearing on parachute journalism, then by developing a modern typology of parachute journalism and making the case for its virtues, and finally by suggesting ways to promote best practice in parachute journalism.

## TWO HISTORICAL PERSPECTIVES

Parachute journalism is typically described as a new development that has thrust itself on the reporting scene. In fact, it has historical antecedents. One of these is the steady advance of air travel technology, which has made parachuting progressively easier. The other is the inherent nature of foreign correspondence, which has made the use of that technology inevitable. Long before they could parachute by plane, reporters overseas jumped off ships, trains, horses, or whatever other transport was available to hop from story to story.

"I do not believe my record will stand for a very great while," Ekins wrote at the end of his around-the-world memoir.[7] He had good reason to draw such a conclusion. None of the previous world records had held up all that long. One of the most famous of all globe-girdling trips was the fictional one in Jules Verne's *Around the World in Eighty Days.* It inspired Nellie Bly's real-life seventy-two-day circumnavigation in 1889–1890.[8] (Both Bly and her somewhat slower competitor, Elizabeth Bisland of *Cosmopolitan,* also wrote books about their trips.) The pace quickened year by year because of the explosion in transportation technology. Verne's hero traveled by different types of carriage: elephant, pony, palanquin, and sledge, as well as paddle-wheel steamer, steamship, and locomotive. Those who followed him in real life used trains that moved faster, automobiles, and then airplanes. The first record breaker to travel by air was a New York theatrical producer in 1913. He completed his trip in thirty-six days, journeying the last forty-mile leg on the wing of a plane.[9]

Commercial air travel, which Ekins introduced into the around-the-world equation, was still out of reach for most Americans when he returned home triumphantly in 1936. But this was about to change. By any measure, air travel became easier and easier in the coming years.

- Airplanes flew to more places. In 1926, the total route mileage of U.S. airlines operating abroad was a mere 152 miles. In 1936 it reached 30,567 miles; on the eve of World War II it had jumped to 53,025 miles; and by 1948 that number more than tripled, hitting 172,177 miles.[10]
- The speed of travel increased. In 1930, the maximum air transportation speed was about 150 miles per hour; by 1950 it was close to 600. The Concorde, introduced in 1976, sped along at 1,350 miles per hour.[11]
- The cost of flying abroad dropped. Between 1937 and 2004, per-mile passenger charges for international travel—expressed in real terms—fell more than 90 percent.[12]
- Accordingly, the number of passengers increased. Just under 109,000 passengers flew abroad on U.S. carriers in 1937; well over 1.3 million did in 1948.[13] This has continued into the present. American Airlines alone carried 18.9 million passengers on international flights in 2004, up from 8.5 million in 1990.[14]

The difference between then and now is vividly illustrated by the 1936 maiden voyage of the *China Clipper*, which carried those three publishers to Asia. The plane was designed for thirty-two passengers. Its first flight on the long 2,410-mile California-to-Hawaii leg carried only eight. The one-way fare to Hong Kong was $950, almost twice the average annual per-capita American income at the time. The trip took six days. Today, in contrast, one can reach Hong Kong in fourteen hours. A round-trip ticket can be purchased for $900, less than the cost of a *one-way* fare in 1936 even without taking inflation into account, and equivalent to only

2.7 percent of average annual per-capita income.[15] A Hong Kong–bound Boeing 747 can carry five hundred passengers.

Little wonder that travel and tourism, which scarcely showed up on nations' balance sheets in the nineteenth century, is one of the primary sources of foreign exchange for industrial and developing countries alike today. It accounts for more than 8 percent of global employment and more than 10 percent of global GDP. And it is still growing. According to 2005 estimates, the annualized growth rate of tourism is 7.3 percent.[16]

No census has totaled the number of trips and total flying miles by reporters going abroad for their news organizations. Were it done, the results would be striking, but not surprising. We know from other evidence the trend is growing, slowly but surely, and involves news media of all sizes. Recently the Pew International Journalism Program surveyed foreign editors of papers whose circulations began at 30,000. It found that thirty-nine of the eighty-one larger newspapers at least occasionally had used parachute journalists; so had seven of the seventy-two editors representing the smaller newspapers in the sample.[17] Not that these editors and reporters always use the term *parachute journalism.* They go abroad because they can, and typically they take the news gathering opportunity almost for granted. When we suggested to one longtime publisher that his 40,000-circulation daily was doing parachute journalism, he commented, "I never thought of it that way."[18]

The increasing ease of the technology of flight is one historical development relevant to the discussion of parachute journalism. The other, which makes its use inevitable, is the traditional foreign correspondents' need to travel from their bureaus. For an example of this, consider Paul Scott Mowrer. Largely forgotten today, he was one of the greatest foreign correspondents of his age, working for one of the greatest foreign news gathering enterprises of all time, the *Chicago Daily News*'s foreign service. When the Pulitzer Prize committee decided to include a new category for "correspondence" in 1929, Mowrer became the first recipient. Mowrer was mostly based in Paris, but was nothing if not peripatetic. As a young reporter, he covered the Balkan conflict in 1912, traveling to the front lines by horse. In the 1920s he penetrated Spanish lines in Morocco to reach Riffian rebel leader Abd-el-krim.[19] Mowrer had no prior experience in either of those countries or many of the others to which he journeyed in search of breaking news. Mowrer's experience is by no means anomalous, as memoir after memoir by traditional foreign correspondents testifies.

Foreign correspondents were expected to move around then for the same compelling reasons they are expected to move around today. No news gathering organization can afford to have someone everywhere that news breaks, and even if any could, it would be pointless to spend the money. No newspaper, magazine, or broadcast can possibly accommodate so much news, nor could readers, listeners, and viewers consume all that news. The prudent approach is to send reporters to stories that have reached a commanding level of attention.

Improved air travel, therefore, has not invented an entirely new approach to news gathering abroad. It has simply made an old urge easier to satisfy. It is abetted by improved communication technology, which makes it easier for a correspondent to report from a strange place. A newspaper correspondent no longer has to find a telegraph office to send a story home. He or she simply calls it in on a personal satellite phone. Television correspondents are even better off, notes Bill Wheatley, recently retired vice president of NBC News. Even in the best of times, it is impossible to have broadcast bureaus everywhere. When a story broke in, say, Morocco, a Rome correspondent might be sent to cover it, but travel was complicated by heavy equipment that took a long time to set up. Then a reporter had to send film to a European city to be processed and edited. From Paris it was shipped to New York. Now correspondents use lightweight video cameras and a portable uplink. A reporter can broadcast live from anywhere. Parachute journalism, Wheatley says, has "certainly expanded the datelines."[20]

For years the number-one jumping-off point has been London. Here exist a relatively large number of stories of interest to Americans. Here American reporters speak the language, so they feel more at home than they do in other foreign posts. Here journalists have traditionally had access to some of the best news transmission facilities. And here one can easily catch a ship, train, or plane to someplace else. London, however, is not the only launching pad. For reporters based in South America, Buenos Aires is a common hub; in East Asia, Tokyo and Hong Kong are. In South Asia, it is Delhi. In sub-Saharan Africa, correspondents typically station themselves and their families in Cape Town or Nairobi.

Correspondents operating in these places travel extensively not only inside their country, but also outside of it. Correspondents in Latin America and Africa, for instance, often operate as regional circuit riders. It is not at all rare for a foreign correspondent to travel more than half the time. "The correspondents in Johannesburg and Cape Town spent varying amounts of time in more distant parts of their territory," observed Swedish social anthropologist Ulf Hannerz in one of the few scholarly studies giving some attention to parachute journalism. "Mike Hanna of CNN described his outfit as a 'travel bureau': since CNN operates on the principle of extreme flexibility, unless he was busy with something in Johannesburg, he might fly in wherever there was a staff shortage, whether it was Jerusalem, Cairo, Moscow, or somewhere else."[21]

Traditional foreign correspondents have learned to be experts at parachuting, developing complicated support systems to help them gather news on the fly. A network like CBS has fixers in Amman and Hong Kong, and crews in Bonn and Johannesburg—all ready to assist an incoming correspondent—as well as reciprocal relations through the European News Exchange, which the network organized in 1992.[22] When it is time for correspondents to move on to a new posting, they typically leave behind for their replacement a Rolodex bulging with the names of translators, drivers, stringers, and other fixers living in the various cities and countries on their regional beat.

Critics of parachute journalism might say that this manner of moving about is not "parachute journalism" at all because professional foreign correspondents with extensive experience in the region are doing it. But this is not the significant distinction it might seem to be. Traditional foreign correspondents who are stationed permanently abroad and roam, and correspondents who fly in from the United States share similar handicaps. Take language ability as an example. Everyone knows that correspondents are better equipped for their task if they know the language of the country they cover. A Hong Kong–based reporter is likely to come with Chinese language training. But what is the chance he or she can speak Korean when sent on a special assignment in Seoul, or Vietnamese when sent to Hanoi? Several years ago, a *Los Angeles Times* reporter commented on the problems of speaking only one foreign language, French, when he divided his time among some fifteen countries. "I can't watch the [television] news in Italy," he said. "I can't listen to the radio in Spain. I can't even read the newspapers in Germany."[23]

Similarly, does it really make a difference if a television or print reporter covering Latin America is based in Miami or Rio? Several years ago Peter Copeland, general manager and editor of the Washington bureau of Scripps, considered posting a Middle East reporter in Rome, a common jumping-off point to the region. He changed his mind because "you could get to the Middle East as quickly from Dulles [airport in Washington] as from Italy."[24] "To a degree," anthropologist Ulf Hannerz has noted, "it probably matters less whether one is in a place for a brief or an extended period, only once or many times. What makes the difference may be simply a cultivated sensitivity toward the passage of time: the medium history."[25]

A second problem with drawing a clear distinction between one kind of parachute journalism and another is that many of the parachute journalists dispatched from the United States have extensive experience both abroad and in the subject matter of the story they are covering—and often this expertise trumps what a permanent foreign correspondent can bring to a story. This misunderstood aspect of parachute journalism as well as others is taken up presently. But first we need a better understanding of the various forms of parachute journalism.

## A TYPOLOGY OF PARACHUTE JOURNALISM

A close look at parachute journalism shows that it has evolved into several forms. We have organized this into a typology:

*The traditional overseas foreign correspondent.* The archetypal foreign correspondent is the one who lives abroad and becomes steeped in foreign affairs. The image endures as the standard by which all foreign reporting is measured. If the number of these correspondents increases, foreign news is deemed to have improved. If they get more airtime or space in their publication, foreign news is all that much better. If the education level of these reporters improves, the quality of their work improves. There is something to be said for each of these propositions. But as noted above, the great majority of foreign correspondents are not planted in one location. They expect to travel from one place to another in search of news. As observed by former CBS senior vice president for news coverage, Marcy

McGinnis, stories rarely happen in a city where a news broadcaster or publisher has a bureau.[26]

*The home-based foreign correspondent.* This is the foreign correspondent with a global beat who lives within driving distance of his or her newspaper. The *Washington Times,* which does not have foreign correspondents abroad on a permanent basis, has three or four reporters on staff who go overseas about once a year. Each one specializes in a certain region and has contacts abroad.[27] The *New Orleans Times-Picayune* closed its Latin American bureau in 2002; one of its former foreign correspondents is based at home and serves as an expert on the region.[28] The *Denver Post* has one seasoned parachutist based at home.[29] The situation is similar for the weekly newsmagazine *U.S. News and World Report.* Its foreign editor describes a core group in its Washington, DC, headquarters "who know the world."[30]

*USA Today* had only three permanent foreign correspondents when we interviewed several of its editors, but it also routinely sends reporters with foreign experience abroad to cover news. A decade of international reporting at the paper, says world editor Eliza Tinsley, has given the organization a stronger base of "people who have been overseas and know what it's like to report—people we trust, who understand what it means to get on the ground, to get a local cell phone, to get a fixer. And with that kind of [background], it's legitimate to use them as firefighters."[31] She describes Barbara Slavin as "our Empress of Rogue States." Slavin specializes in Iran, North Korea, and Libya. When we interviewed him, the new senior assignment editor envisioned having one or two of these home-based correspondents "on the road all the time."[32]

One of the few scholars to chronicle the emergence of the home-based foreign correspondent is former correspondent and journalism professor Donald Shanor. He notes that because of the Internet and other worldwide news sources, it is possible to keep up with events without being abroad permanently.[33] Nevertheless, the home-based foreign correspondent is not new. The *Minneapolis Tribune,* acquired by John Cowles Sr. in the 1930s, had a strong interest in foreign news under his ownership and sent reporters overseas "not so much to cover breaking news—the paper's wire services provided enough of that—as to explore and research situations that might underlie future crises and wars."[34]

*The beat reporter abroad.* Arguably the most important international development of the twentieth century was not the rise of communism, or its fall, or even the threat of nuclear weapons. It was the rise of global interdependence, which is manifested economically, culturally, and environmentally. The impact on Americans is tangible daily, and this has given rise to more reporting that ties events abroad to conditions on Main Streets in cities and towns. Much of this reporting can be done locally, as various projects have shown.[35] But cheap airfares have allowed local reporters to trace local stories to their foreign sources and do their own foreign reporting on big topics that seem important even if the ties to home are not direct.

Some local newspapers have been remarkably aggressive in going after general stories abroad (the *Anniston [Alabama] Star* has been notable in this regard). A number have sent reporters on special assignments in recent years. The *Cincinnati Post* sent a reporter who was a Vietnam War veteran to tell the story of "Vietnam today";[36] the *Arkansas Democrat-Gazette* did a special section on Afghanistan a year after the U.S. invasion;[37] the *Hartford Courant* sent a French-speaking reporter to assess the climate in Paris before the Iraq War;[38] the *Birmingham News* sent a reporter to Iraq (he was "a world traveler, so he didn't need a lot of handholding").[39]

By far the greatest emphasis, however, is on stories with local connections. A few examples suggest how significant and far-reaching this can be: When the *Seattle Times* did a series on the global economy and outsourcing, a technology reporter went to India to cover the expansion of companies like Redmond-based Microsoft, and a business reporter went to Peru, where Del Monte had moved its asparagus-growing operation after years of running it in Washington.[40] The *Daily Herald,* a suburban newspaper in Chicago, has sent reporters to the Philippines, Poland, India, and Mexico to trace ties to local immigrant groups.[41] The *Eugene Register-Guard* dispatched a reporter to Kenya to do a story on a local nurse practitioner who ran an AIDS orphanage there and sent a coastal reporter to Japan to see what its approaches were to tsunami preparedness.[42] The *Wilmington Morning Star* sent its business editor to Cuba to cover an agricultural exhibition, which was particularly salient to North Carolina because of its turkey industry—which falls outside of Cuba's trade embargo.[43] When we were doing our interviews, the *Lincoln Journal-*

*Star* was planning to send a reporter and photographer to Afghanistan to cover a childhood education program headed by a University of Nebraska professor.[44] The focus in all these cases was clearly local. "We are definitely not there for the big picture," the editor of the *Honolulu Advertiser* said of the paper's reporter in Iraq.[45]

In many cases these journalists will work under "embedded" conditions. This would include, for example, the *Cincinnati Post* reporter who accompanied a mission group to tsunami-devastated Indonesia; the *New Haven Register* reporter who went to Europe with a local basketball team; the business reporter from the *Tulsa World* who traveled to Saudi Arabia with a local oil company; and the education reporter for the *Des Moines Register* who traveled to Mexico with a team of school administrators to cover their participation in a federal program designed to help Iowa's immigrant students perform better in school.[46]

*The vacationing reporter.* Critics often portray parachute journalism as the work of amateurs—and the vacationing reporter seems the closest to this, although their work can be valuable. In some cases a staffer will use vacation time to participate in an overseas medical mission and then write a story about the experience; other times an enterprising reporter studying abroad or vacationing simply wants to track down a story overseas that interests him or her. A *Cincinnati Post* reporter on vacation in London did a story on a new play that happened to have been written by Cincinnati's former mayor, Jerry Springer.[47] A *Bakersfield Californian* reporter used a vacation in Israel and Singapore to write "terror-related pieces."[48] A *Hartford Courant* reporter accompanied a team of Connecticut climbers scaling Everest.[49] And at the *Lincoln Journal-Star,* a sweet journalistic prize comes with the Employee of the Year award: the reporter is allowed to do a story of his or her choice and may go aboard to do it. Reporters have traveled to Guatemala, Germany, and India, among other places.[50]

In some cases, a reporter happens to be on hand when a story breaks. A *Long Beach Press-Telegram* sports writer was vacationing in the Soviet Union when Chernobyl blew up. He went to the scene and filed stories for a week.[51] Reporters may get some support from their newspaper in the form of expenses or payment for the story or an extra day or two of vacation.

## MISCONCEPTIONS AND VIRTUES

The work of these four types of parachute foreign correspondents is not of a uniform quality, and as we have noted, the use of parachute journalists to eliminate permanent correspondents based abroad is tragic for American consumers of news. Nevertheless much of the broad, undifferentiated criticism of parachute journalism is misconceived. Parachute journalism has demonstrable virtues.

### Virtue 1: Parachute Journalism Can Be Additive

Contrary to the common misconception, parachute journalism is not always a substitute for permanent foreign correspondents. Small- and medium-sized newspapers that previously could never have hoped to have a reporter overseas under any circumstances are able to cover stories of special interest. Above we listed some examples of special missions by small- and medium-sized newspapers. But larger newspapers use parachute journalists to increase coverage, not cut back on it.

*Chicago Tribune* foreign editor Timothy J. McNulty has eleven full-time foreign correspondents. That is a respectable sized corps of foreign correspondents, but they are thinly stretched nonetheless. The best way to add personnel, he says, is to draw on the metro desk, which has over a hundred reporters who can be brought in on stories. Virtually any story may turn into a foreign assignment. When a former Chicago Bulls basketball player was lost in the South Pacific, a metro reporter was sent out to French Polynesia. Asked how often he draws from staff elsewhere on the paper, McNulty started to make a count and then gave up. "There are examples all the time."[52] David Hoffman, assistant managing editor for foreign news at the *Washington Post,* expressed a similar sentiment: "In a world that is so big and complex, I don't cover it all. I am always looking to expand my resources."[53] The best way is to throw domestic reporters into stories.

### Virtue 2: Parachute Journalism Can Bring More, Not Less, Expertise

A related virtue runs counter to common misconceptions that parachute foreign correspondents come with no real expertise. The contrary is often true. The *Washington Post,* the *Boston Globe,* and the *Chicago Tribune* sent religion reporters to Rome to cover the death of Pope John Paul II and the election of a successor because these reporters were better equipped than their permanent correspondents to cover this complex story. Similarly, the *Boston Globe* sent a Pentagon reporter to an arms control meeting in Oslo, and the *Chicago Tribune* sent a music reporter to Cuba.[54] The question is not so much whether the parachute journalist has the necessary background, says the *Washington Post's* David Hoffman; it is whether the resident correspondent's knowledge is "too limited."[55]

In many cases, parachute reporters are teamed with resident reporters to master a story. Doing a story on priests who have been convicted of crimes and fled the country, the *Dallas Morning News* sent two reporters to the Philippines, Africa, central Europe, and elsewhere.[56] Only one had been a foreign correspondent previously. When the *Chicago Tribune* set out to do a story on deportations of immigrants with visa problems on night flights to Pakistan, a correspondent based in New Delhi, a reporter with the Washington bureau, and a metro reporter became involved.[57]

This ability to use travel to marshal expertise is prompting editors to rethink what it means to be a foreign correspondent. The *Boston Globe* put a reporter with special expertise in Cape Town. He covered the normal run of foreign news but also paid special attention to medical issues not only within the region but outside it. He was seconded to Indonesia to cover the aftermath of the Tsunami.[58] The *Washington Post's* Hoffman envisions a new way of marshaling expertise. "My vision," he said, "is to break the bonds of individual geographic locations and create transnational international reporting."[59] Specifically, he is contemplating the establishment of four or five global beats on such topics as democracy and freedom, or the environment. The correspondents might be based abroad or in the United States. Either way, considerable parachuting would be involved.

### Virtue 3: Parachute Journalism Can Help Reporters Connect with Readers, Viewers, and Listeners

Among the perennial questions about foreign news coverage that are never adequately resolved is how much time a reporter should remain abroad, let alone in one country. A reporter who stays overseas a long time gains expertise but loses touch with the audience for which he or she is writing.

A famous story about Colonel Robert McCormick, who idiosyncratically ran the *Chicago Tribune* for decades, illustrates this concern. After questioning a group of would-be correspondents about their language ability, he selected the one who spoke no second tongue at all. "I don't want my fine young Americans boys ruined by these damn foreigners," explained McCormick.[60] The concern reappears in a comment from a *New York Times* foreign editor who said the best policy is to keep correspondents in one place no longer than four or five years; otherwise they "have the tendency to go native."[61]

While substantial experience overseas is a significant plus in foreign correspondents, all students of foreign news coverage must admit that news media face a considerable challenge in interesting audiences in foreign news. Reporters who understand that audience, who can relate foreign stories to their readers, viewers, and listeners, are important not only because such reporting is in itself useful but also because it can elevate the importance of foreign affairs among Americans. Home-based journalists can excel at this, for they have an edge in identifying tangible links between foreign stories and local circumstances. It can be argued that they create a more attentive audience for traditional foreign correspondence, which may be essential to its future robustness.

One bit of evidence to suggest that parachute journalism increases public interest in foreign news comes from surveys done when the *Minneapolis Tribune* was practicing the art. In 1944, the first year of the *Tribune's* Minnesota Poll, two-thirds of respondents favored membership in a United Nations–type organization. The next year the figure rose to four out of five, and by 1953, a full 70 percent of Minnesotans favored continued UN membership. This is extraordinary for a state with an isolationist tradition. "Because most of the Minneapolis reporting was done not by Washington or overseas-based correspondents," observed Charles Bailey, once the *Tribune's* editor, "but rather by reporters who lived and worked among their readers, it was easier and more natural to keep in mind such questions of relevance."[62]

**Virtue 4: Parachute Journalism Gives Media New Management Tools**

Parachute journalism provides flexibility that did not exist before. This is not only a matter of allowing editors to get more reporters into the field.

Newspapers can use parachute assignments to see if a reporter is suited for a permanent foreign assignment. The *Chicago Tribune* reporter who was sent to French Polynesia on the lost basketball player story was later posted abroad permanently. He had done a good job.[63] Parachute assignments also serve as a recruiting tool. Many first-class reporters gravitate to newspapers that will give them a foreign assignment at some point. While a parachute assignment is not the same thing as a permanent overseas posting, it is better than nothing—and given the rise of two-career couples, which makes it more difficult for families to uproot themselves for two or three years, it may even be preferable for some individuals. The possibility of parachute assignments can also make it easier to retain a reporter who has come back from an overseas bureau and otherwise might leave for a newspaper that will send him abroad again.

Parachute journalism also takes advantage of the evolution of news media into sprawling corporations. Without a doubt, the corporatization of news media presents problems. Inevitably the foreign staffs of the *Los Angeles Times* and the *Chicago Tribune,* now owned by one company, will be amalgamated in some way. While this book was going to press, the Tribune Company eliminated the fabled foreign service run by another one of its newspapers, the *Baltimore Sun,* and the *Boston Globe,* owned by the *New York Times,* eliminated all of its foreign bureaus. But large organizations also offer economies of scale that work in favor of their smaller newspapers, whose budgets could not possibly sustain even one permanent foreign correspondent. If a smaller Knight Ridder newspaper wants to send a reporter abroad on a short-term assignment, it can turn to the parent company or one of its larger newspapers for assistance in lining up translators and other helpers overseas. The *Detroit News,* for instance, is part of the Gannett newspaper group. "If they're going to Baghdad—and this has happened—the *Detroit News* will call and say, 'Can we use your hotel room?'" said *USA Today*'s Eliza Tinsley, who responded, "Sure. You know you can camp out on the floor [but] the beds are taken. And [you can] use our fixers."[64]

## WHERE DO WE GO FROM HERE?

As anyone familiar with journalism history knows, foreign news coverage is precarious. Paul Scott Mowrer's *Chicago Daily News* virtually invented the concept of high-quality foreign news, but that did not save it from extinction. Its foreign news service disappeared before the newspaper did in 1978. The pressures that work against foreign news show no sign of abating. Foreign news is extremely important, given American involvement in world events, but that news also involves high costs and enjoys a relatively small audience except in times of crisis.

This is not to excuse news media from doing a better job of traditional foreign news gathering. Permanent foreign correspondents posted abroad are essential for news media with the capacity to undertake such initiatives. *USA Today,* with only a handful of overseas reporters, falls far short of what should be expected of a daily that advertises itself as a national newspaper and circulates extensively abroad. And how can anyone have much regard for the daily *Washington Times* in the nation's capital when it uses only parachute journalists?

Nevertheless, parachute journalists can be an answer to legitimate problems of cost—and can improve coverage even when money is no object. Small- and mid-sized newspapers that could never before afford to send anyone abroad can do so now. For larger newspapers, parachute correspondents add muscle and expertise to foreign coverage. For any newspaper, parachute journalists who have their customers clearly in mind can help enlarge the audience for foreign news.

Without a doubt, parachute journalism has limitations. Ekins's comment at the end of his globe-girdling race suggests a sort of worst-case outcome: "Observations had to be fleeting. Contacts with humankind had to be few. Impressions had to be, and were, superficial."[65] But there is more to parachute journalism than hit-and-run reporting. Taking advantage of these potentials requires a new mind-set. Many editors are finding ways to use parachute journalists effectively either to supplement regular overseas reporting or to mine the local angle. Much of this is ad hoc, however, and not part of a large strategy. Media companies need to think more about best practices: how to design assignments, how to prepare reporters ahead of time, how to support them in the field. Large news organizations with considerable experience abroad have an advantage in this. Smaller ones with no foreign news tradition need help. Significantly, in our interviews, editors with many newspapers were vague about how they prepared reporters for one-shot assignments abroad.

Some initiatives point in the right direction. In the run-up to the Iraq War, special training camps prepared would-be war correspondents for combat assignments. In 2002, the International Center for Journalists (ICFJ), the World Affairs Councils of America, and the Newspaper Association Managers launched a program sending reporters from medium and small media markets in search of foreign links to their communities.[66] The International Reporting Project based at Johns Hopkins University sends American reporters from around the country abroad on short-term assignments. The philosophy, says John Shidlovsky, the project director, is to build up foreign capabilities on news staff so home-based journalists can do a week abroad here or there as needed.[67]

ICFJ and other organizations with extensive media contacts abroad can help parachute correspondents by putting them in touch with fixers and experts abroad. The American Press Institute could be holding seminars on parachute journalism, as could large news media companies when they bring editors together. The subject should be on the agenda of the annual meeting of the American Society of Newspaper Editors.

Scholars need to do more as well. They need to think more about parachute journalism's various iterations. (Indeed, we hope scholars will refine our early rough typology outlined above.) They need to study more carefully the content of such reporting. We have been struck by how little has been done by academics on this subject. This is a field ripe for study and clarification.

Parachute journalism is not a new concept, but it is growing in frequency and complexity. Unfortunately, it has been portrayed just as simplistically as its critics accuse parachute journalists of portraying the news. Technology is neutral; how we use it is up to us. Understanding how to make the best use of jet travel is far better for journalism and the public than dismissing parachute journalism out of hand. Happy landings are possible.

NOTES

1. H. R. Ekins, "A Reporter Aloft," in *We Cover the World: By Sixteen Foreign Correspondents,* ed. Eugene Lyons (New York: Harcourt, Brace, 1937), 371–88; see also H. R. Ekins, *Around the World in Eighteen Days and How to Do It* (New York: Longmans, Green, 1936).

2. "Flier Sets Records in Tasman Sea Hop," *New York Times,* October 17, 1936.

3. Robert L. Gandt, *China Clipper: The Age of the Great Flying Boats* (Annapolis: Naval Institute Press, 1991), 107–9; T. A. Heppenheimer, *Turbulent Skies: A History of Commercial Aviation* (New York: John Wiley, 1995), 71. The newspaper executives on the *China Clipper* were—in addition to Roy W. Howard, of Scripps-Howard—Paul Patterson, president of the *Baltimore Sun,* and Amon Carter, publisher of the *Fort Worth Star-Telegram.*

4. A typical estimate is that it costs $250,000 annually to station a newspaper foreign correspondent abroad. And for a broadcast journalist to be there the cost is much higher.

5. Mort Rosenblum, *Coups and Earthquakes: Reporting the World for America* (New York: Harper and Row, 1979), 11.

6. Tom Fenton, *Bad News: The Decline of Reporting, the Business of News, and the Danger to Us All* (New York: Regan Books, 2005), 63. For other examples of such criticism, see Namrata Savoor, "'Parachute Journalism' Hurts World News Overseas," Freedom Forum Web site, May 30, 2001, http://www.freedomforum.org/templates/document.asp?documentID=14034; Tunku Varadarajan, "Parachute Journalism Redux," *Wall Street Journal,* November 12, 2001.

7. Ekins, *Around the World in Eighteen Days,* 167.

8. Brooke Kroeger, *Nellie Bly: Daredevil, Reporter, Feminist* (New York: Times Books, 1994); Jason Marks, *Around the World in 72 Days: The Race between Pulitzer's Nellie Bly and* Cosmopolitan*'s Elizabeth Bisland* (New York: Gemittarius, 1993).

9. Carroll V. Glines, *Round-the-World Flights* (New York: Van Nostrand Reinhold, 1982), 12–13.

10. Rudolf Modley, *Aviation Facts and Figures, 1945* (New York: McGraw-Hill, 1945), 71–72; G. Lloyd Wilson and Leslie A. Bryan, *Air Transportation* (New York: Prentice-Hall, 1949), 205.

11. Jerome C. Hunsaker, *Aeronautics* (Oxford, UK: Oxford University Press, 1952), 30; John C. Spychalski, "Transportation," in *The Columbia History of the 20th Century,* ed. Richard W. Bulliet (New York: Columbia University Press, 1998), 411.

12. This calculation is based on data from the Air Transport Association, http://www.airlines.org/economics/finance/PaPricesYield.htm, as of July 18, 2005.

13. Wilson and Bryan, 267.

14. "Airlines Find Help in Flights Overseas," *Wall Street Journal,* July 5, 2005.

15. Gandt, 107–9; per-capita income data from *Statistical Abstract of the United States* (Washington, DC: U.S. Department of Commerce, 1993), 445.

16. World Travel and Tourism Council, *World Travel and Tourism: Sowing the Seeds of Growth,* World Travel and Tourism Council Report no. 6 (London: World Travel and Tourism Council, 2005), 36–37.

17. Dwight L. Morris and Associates, *America and the World: The Impact of Sep-*

*tember 11 on U.S. Coverage of International News* (Washington, DC: Pew International Journalism Program, 2002), 4.

18. Joe D. Smith, interview by John Maxwell Hamilton, 2004.

19. Paul Scott Mowrer, *House of Europe* (Boston: Houghton Mifflin, 1945), 202–3, 349.

20. Bill Wheatley, interview by Hamilton, New York, NY, May 3, 2005.

21. Ulf Hannerz, *Foreign News: Exploring the World of Foreign Correspondents* (Chicago: University of Chicago Press, 2004), 63–64.

22. Marcy McGinnis, interview by Hamilton, New York, NY, May 19, 2004.

23. David Shaw, "Foreign Correspondents: It's On-the-Job Training," *Los Angeles Times,* July 2 1986.

24. Peter Copeland, interview by Hamilton, Washington, DC, November 5, 2003.

25. Hannerz, 230.

26. McGinnis interview.

27. David Jones, telephone interview by Emily Erickson, December 3, 2004.

28. Jed Horne, telephone interview by Erickson, February 11, 2005.

29. Bruce Finley, telephone interview by Erickson, November 12, 2004.

30. Terry Atlas, interview by Hamilton, Washington, DC, February 11, 2004.

31. Eliza Tinsley, interview by Hamilton, Roslyn, VA, February 2, 2005.

32. James Cox, interview by Hamilton, Roslyn, VA, February 2, 2005.

33. Donald R. Shanor, *News from Abroad* (New York: Columbia University Press, 2003), 45–46.

34. Charles W. Bailey, "Foreign Policy and the Provincial Press," in *The Media and Foreign Policy,* ed. Simon Serfaty (New York: St. Martin's, 1990), 185.

35. John Maxwell Hamilton, *Main Street America and the Third World,* 2nd ed. (Cabin John, MD: Seven Locks Press, 1988); American Society of Newspaper Editors, *Bringing the World Home: Showing Readers Their Global Connections* (American Society of Newspaper Editors, 1999), http://www.asne.org/index.cfm?ID=2569.

36. Mark Neikirk, telephone interview by Erickson, June 2, 2005.

37. Bob Wigginton, telephone interview by Erickson, June 25, 2004.

38. Bernie Davidow, telephone interview by Erickson, January 11, 2005.

39. Stan Diel, telephone interview by Erickson, June 25, 2004.

40. Jim Simon, telephone interview by Erickson, January 17, 2005.

41. Teresa Schmedding, telephone interview by Erickson, January 16, 2005.

42. Kevin Miller, telephone interview by Erickson, June 2, 2005.

43. Bonnie Eksten, telephone interview by Erickson, June 2, 2005.

44. Shelly Kulhanek, telephone interview by Erickson, June 6, 2005.

45. Steve Patrenik, telephone interview by Erickson, October 12, 2004.

46. Neikirk interview.

47. Ibid.

48. Logan Molen, telephone interview by Erickson, June 24, 2004.

49. Davidow interview.

50. Shelly Kulhanek, telephone interview by Erickson, June 6, 2005.

51. John Futch, telephone interview by Erickson, June 28, 2004.

52. Timothy McNulty, telephone interview by Hamilton, March 23, 2005.

53. David Hoffman, interview by Hamilton, Washington, DC, April 14, 2005.

54. McNulty interview; John Yemma, telephone interview by Hamilton, March 10, 2005.

55. Hoffman interview.

56. Bob Mong, interview by Hamilton, Baton Rouge, LA, April 2, 2005.

57. McNulty interview.

58. Yemma interview.

59. Hoffman interview.

60. Richard Norton Smith, *The Colonel: The Life and Legend of Robert R. McCormick, 1880–1955* (Boston: Houghton Mifflin, 1997), 300.

61. Roger Cohen, interview by Hamilton, New York, NY, September 17, 2003.

62. Bailey, 185.

63. McNulty interview.

64. Tinsley interview.

65. Ekins, *Around the World in Eighteen Days,* 381.

66. John Maxwell Hamilton and Eric Jenner, "Redefining Foreign Correspondence," *Journalism: Theory, Practice and Criticism* 5 (August 2004): 308.

67. John Shidlovsky, interview by Hamilton, Washington, DC, May 10, 2005.

# 8

# THE REAL-TIME CHALLENGE

## *Speed and the Integrity of International News Coverage*

PHILIP SEIB

Although real-time news technologies are still evolving, they have already transformed not only the processes of journalism, but also the effects of journalism on the public and policy makers. A problem underlying this transformation—and one that weakens its chances to stimulate constructive change—is journalism's overemphasis on speed, which has been given a steroidlike boost by new technology. This chapter will consider the conventional wisdom that "faster is better" and will make the case that this flawed journalistic standard is particularly harmful when applied to international news coverage.

One rationale behind high-speed journalism is that in an era of terrorism and dispersed weapons of mass destruction, people want to receive information as quickly as possible. But the complexity of international relations today makes such an approach error-prone, and the resulting news product is likely to be simplistic to the point of being misleading.

### EFFECTS ON THE NEWS PROCESS

Real-time international news coverage became an important journalistic tool when Edward R. Murrow broadcast live from England during the first years of World War II. He stood on London rooftops as German bombs exploded and British antiaircraft fire ripped the night, delivering the action, as it happened, into America's living rooms. Archibald MacLeish said of Murrow's radio reporting:

> You destroyed in the minds of many men and women in this country the superstition that what is done beyond three thousand miles of water is not really done at all; the ignorant superstition that violence

> and lies and murder on another continent are not violence and lies and murder here. . . . You burned the city of London in our houses and we felt the flames that burned it. You laid the dead of London at our doors and we knew the dead were our dead—were all men's dead—were mankind's dead—and ours.[1]

The vividness and drama of live reporting enhanced the power of Murrow's coverage and enthralled his listeners. They felt themselves to be at the scene and part of the story, something that the print media could not match. The coverage also helped change Americans' opinions about their country's foreign policy. Murrow's real-time descriptions of London under siege nudged Americans away from isolationism and toward a more assertive policy of aiding Britain.

The real-time style pioneered by Murrow became a distinguishing characteristic of radio and then television and was used in the competition for audience. "First with the news" was promoted as an essential attribute. That may seem perfectly legitimate, but emphasis on going live is susceptible to being abused. In international and domestic journalism, there is a tendency to rely on "live for the sake of live," on the assumption that the intrinsic drama of real-time reporting is more important than the newsworthiness of what is being covered. The classic example is a local television station covering an inconsequential car chase for hours or presenting a live shot from a location where events being recounted took place hours before. It would be somewhat comforting to be able to dismiss such nonsense as being "just local TV news," but unfortunately the practice has infected coverage by news organizations that should know better and has extended into reports about important international news stories.

One reason that live coverage has become so common—even in inappropriate circumstances—is the development of technology that makes real-time reporting easier and less expensive. The 1991 Gulf War was the first television-era conflict to have been covered largely in real time. During that war, mobile live reporting by satellite involved a transmitter so large that a truck was needed to carry it, and the gear cost about $100,000. By 2003, reporters covering the Iraq War were using briefcase-size satellite transmission kits that weighed approximately fifteen pounds and cost about $20,000. Television networks were even able to report while on the move with forces racing through Iraq. CNN used a videophone connected to an enclosed antenna on a gyroscope-controlled platform that kept it pointed toward the satellite, and NBC reporters sent their signal to a truck that trailed them by two miles and carried a gyroscope-aided satellite dish encased in a dome. The next generation of field reporting equipment is expected to be the size of a laptop computer, weigh five pounds, and cost $3,000.[2] Increased mobility and decreased cost mean that more news organizations and even freelance journalists will be able to generate real-time reporting, and "going live" will be the standard, not the exception.

In a technological sense, that certainly is progress. But does better technology necessarily lead to better journalism? As coverage of the Iraq War indicated, news organizations can fall into two traps:

- Providing premature and incomplete reports to feed the presumed public appetite for constant real-time coverage.
- Giving precedence to live reports at the expense of background stories that are essential in providing the depth and context that solid coverage of wars and other international news events requires.

By its nature, news involves the description of incomplete events. No news organization, in any medium, will wait to report a story until it can be told in its entirety. In part, "news" as a product consists of updating an established story line in response to the question that always hangs in the air, "What's the latest?" But as a matter of professional responsibility, the craft of journalism requires that this ongoing updating should be solidly substantiated and involve a minimum of speculation. Real-time pressures work against this standard, as is illustrated in coverage of the thrusts and parries of warfare, diplomacy, and other aspects of international relations that take time to reach their eventual outcomes. News organizations are reluctant to say, "Let's wait a bit to see how this is developing," because the increasingly prevalent standard is immediacy rather than accuracy. Individual journalists have to give up "think time" because they are expected to deliver live reports as events unfold around them.[3]

During the Iraq War, wrote Andrew Hoskins, "the journalists closest to the heart of battle itself ironically contributed mostly narrow and decontextualized snapshots of the war."[4] This was due partly to the limited range of vision that was available to embedded reporters and others nearest the action, but the demand for coverage also required these reporters to steadily feed the real-time beast whatever morsels were available. This was a function of the emphasis by news organizations—especially those of the major television networks—on providing the most dramatic reporting, even if it was not the most enlightening. Pentagon officials referred to the narrowness of embedded reporters' perspective as viewing the war through soda straws.[5] Combat is only part of war; beyond the chaos of the battlefield are politics, diplomacy, economics, and other drier sciences. Much news coverage of the Iraq War, particularly television's, was in too much of a rush to address this breadth properly.[6]

Another medium that has proved problematic for news organizations is the Internet. News Web sites are the latest manifestation of real-time news delivery. They can be updated instantly and constantly, like a newspaper remaking its front page every few minutes. The Web means news can be real-time all the time, which should inspire caution as well as ambition because it changes the culture of the newsroom. Journalism has traditionally involved a metronomic process of gathering, investigating, verifying, gathering some more, corroborating further, and scrutinizing, all to produce a product at regular intervals. Web-based news changes that drastically, leading some newspaper editors to observe that they now operate on the same schedule as CNN does.[7] That means that more news will be delivered more quickly, but it raises the question of whether that is necessarily a good thing.

Technology per se should not be blamed for speed-related errors in journalism; humans, not gadgets, are making the mistakes. The controversy surrounding *Newsweek*'s 2005 story about disrespectful treatment of the Koran by U.S. military interrogators illustrates that premature reporting can have lethal consequences, as the story spurred rioting in Afghanistan and elsewhere in which more than a dozen people were killed and upwards of a hundred were injured. As *Washington Post* ombudsman Michael Getler observed, *Newsweek* stated that the incident mentioned in the magazine "was 'expected' to be included in an upcoming report. That's like saying somebody is 'expected' to be indicted or found guilty, and journalists simply don't do that. Here is one more case where it was definitely worth waiting to make sure of the facts."[8]

Process matters in journalism. The basic story about prisoner abuse that *Newsweek* was reporting was true, but by neglecting process—in this case, inadequately corroborating the information—the magazine's journalists undercut the credibility of their reporting.

The *Newsweek* incident demonstrates the corrupting influence of wanting to get a story first rather than wanting to first get it right. Its reverberations underscore the exceptionally high stakes that may be involved when an error resulting from moving too quickly sweeps across the globe. The lesson for news professionals is obvious, but there is little evidence that it will be heeded.

## EFFECTS ON NEWS CONSUMERS

The public has become conditioned to expect a steady diet of "breaking" news. Given the assiduous self-promotion by news organizations about their abilities to deliver real-time news, it is not surprising that their audience tends to equate "most recent" with "most important." The illogic of this equation has not been a deterrent to its popular acceptance.

Traditional news media stress their real-time capabilities as one aspect of their increasingly desperate struggle to halt the diffusion of news audiences that has left industry giants looking puny. Since the 1990s, news consumers have found many new ways to get real-time information.

*Expanded television choices.* With its coverage of the 1991 Gulf War, CNN's all-news format established news-on-demand as a potent rival to broadcast networks' scheduled newscasts. Live continuous coverage was no longer exceptional; it had become everyday fare and changed the audience's expectations and viewing habits.

*News Web sites.* Stories of secondary magnitude—highly significant but not so important as to merit nonstop television coverage—can be monitored on line. The *New York Times, Washington Post,* and BBC Web sites are among those that are updated almost minute by minute when the story merits it. The Web also encourages global news sampling because it is so easy to tap into coverage by Lebanon's *Daily Star,* Japan's *Asahi Shimbun,* France's *Le Monde,* and countless others. Whether because of dissatisfaction with usual news sources or simply out of curiosity, American visits to the BBC World Service's Web site increased 158 percent during the first weeks of the Iraq War.[9] As broadband capability and media convergence have increased, many sites have begun to offer high-quality video and audio as well as text, giving news consumers yet another reason to choose the Internet as a news provider.

*Unmediated media.* The Internet makes it easy for news consumers to act as quasi journalists, gathering information from primary sources rather than waiting for news organizations to collect and package it. The White House, the Iranian Ministry of Foreign Affairs, NATO, OPEC, Doctors without Borders, and a global roster of other official and unofficial sources are just a mouse click away, changing the relationship between public and news media. For Americans, this is a useful supplement to other information sources. For many others throughout the world, in countries where news media lack freedom and substance, this is a far more significant advance because it provides unprecedented access to unscreened information (although some governments have so far been able to obstruct Web access). Despite the allure of intellectual independence, this approach may leave the public susceptible to propaganda or disinformation that professional journalists might challenge and filter out.

These venues—including some of the government sources—often feature real-time reports. This contributes to growing public expectations and demand for real-time coverage, as was seen in audience movement to providers of such coverage. For U.S. cable television during the initial stages of the Iraq War, the average number of daily viewers of MSNBC and CNN increased more than 300 percent, and the Fox News audience grew 288 percent.[10]

Meanwhile, the big three broadcast networks watched viewers depart. Early in the war, CBS and ABC together lost approximately two million viewers, a combined 10 percent, while NBC, the only one of the big three to also have a cable news operation, saw a slight increase. Normally, the networks gain audience during a war or other big news event. Never before while such a major story has unfolded have the big three evening newscasts seen their viewership decline. One reason may have been that the cable networks provided real-time coverage whenever events warranted, while the broadcast networks were more locked into their schedules. Also, according to CBS News president Andrew Heyward, the role of the network anchor is not as crucial in reporting a war as it is during singular events such as the 9/11 attacks. He said that the Iraq conflict "was a reporters' war, not an anchor war," which let the cable networks compete more effectively.[11]

More recent evidence of the audience appeal of real-time coverage could be seen in the Middle East in spring 2005. Web sites carried real-time video and audio from Martyr's Square in Beirut as hundreds of thousands of Lebanese demanded that Syria withdraw from their country. In addition to appearing on already existing news sites, this coverage was also provided by ad hoc purveyors, helping to galvanize Lebanon's "Cedar Revolution" and stirring interest in the possibility of political change elsewhere in the Arab world.

Overall, the global audience for international news is increasingly accustomed to finding real-time coverage available through various media. In the Middle East, Al Jazeera is just the best known of more than 150 regional satellite television stations. In Latin America, a regional network, Telesur, has been endorsed by governments as an alternative to Spanish-language news providers from outside the region. This proliferation of indigenous media is likely to continue throughout the world as more satellites are launched, costs decline, and the satellite television audience expands. Equally explosive will be the growth of Internet use by news consumers. Real-time news coverage will be an important asset in the competition for audience, and as different media become available on handheld devices as well as other instruments, a substantial part of the public will be able to monitor real-time news.

But will this technological triumph ensure that people are better informed? Probably not, unless content takes primacy over speed. The ups and downs of television illustrate that this is by no means a certainty; any medium is susceptible to becoming little more than a repository for clutter.

## EFFECTS ON POLICY MAKING

Information affects opinion; opinion affects politics; politics affects policy. That boiled-down formulation begins to explain why policy makers are sensitive to news coverage. Public attitudes about foreign affairs are particularly susceptible to news media influence because most people have only a shallow pool of personal knowledge or experience from which to draw about such matters and so depend more on news reports. Political leaders have long felt the sting of media criticism of their foreign policies: George

Washington's embrace of neutrality was attacked by the *National Gazette,* William McKinley was shoved along toward the Spanish-American War by newspapers owned by William Randolph Hearst and Joseph Pulitzer, and many other shapers of policy have found themselves pressured by news-influenced public opinion.[12]

Television changed the dynamic between foreign policy makers and the American public by bringing dramatic images of war and other disasters into the living room. Even more than Murrow's London broadcasts, this coverage affected many in the audience. During the Vietnam War, wrote the *New Yorker*'s Michael Arlen, television viewers watched "real men get shot at, real men (our surrogates, in fact) get killed and wounded."[13] The vividness of this coverage made it imperative for government officials to better explain and defend their policies. None of the reporting from Vietnam was in real time, but as news coverage increased in speed and quantity, it was affecting awareness and expectations.

The 1991 Gulf War can be designated as the birth of the era of real-time international news coverage because much of this continuing story was reported live or after only brief delay. Later that year, the intersection of coverage and policy making could be seen during the short-lived coup d'etat in the Soviet Union. The TASS news agency initially reported that Soviet president Mikhail Gorbachev was stepping down due to "ill health," although he actually had been placed under house arrest. Brent Scowcroft, national security adviser to President George H. W. Bush, received his early information about events by watching CNN, and General Colin Powell, chairman of the Joint Chiefs of Staff, "kept one eye on CNN and another on intelligence reports that were still flowing in."[14]

In Moscow, opposition to the coup was coming together, led by Boris Yeltsin, and large numbers of people were on the streets to protest the seizure of power. A Bush administration official (who declined to be named) told the *Washington Post* that "his first consideration on hearing about the coup was not how to cable instructions on U.S. reaction to American diplomats, but how to get a statement on CNN that would help shape the response of all the allies. 'Diplomatic communications just can't keep up with CNN,' he said. . . . 'We had also sent a signal to Yeltsin and the people on the street that we are going to work toward elimination [of the coup], that we are with you. Yeltsin finds out that Bush is on his side—he finds out publicly before we could even get a message to him.'" Secretary of State James A. Baker later wrote that Bush "had used the fastest source available for getting a message to Moscow—CNN."[15]

This case is not unique. Policy makers use the news media to gather information and to send messages, and the real-time capabilities of news organizations are invaluable in both instances. Reliance by officials on news reports underscores the importance of maintaining high standards of accuracy. Although it is not likely that a government would take drastic action based solely on a news story, news organizations may run well ahead of official sources and so may, on occasion, be the de facto primary providers of information.

Policy makers must deal with this aspect of news media influence on another level: responding to how the public reacts to news coverage. This is the so-called CNN effect, which is generally defined as news coverage—particularly real-time television reports—shaping policy. This concept should be approached carefully, recognizing the important distinction between news coverage *influencing* policy and news coverage *determining* policy. The former does happen to varying degrees, but only a government wholly without anchor would find its policies truly driven by news reports. Stephen Badsey accurately appraised this process when he wrote that "although the CNN effect may happen, it is unusual, unpredictable, and part of a complex relationship of factors."[16] That said, no policy maker with any sense should discount the potency of news reports, particularly when the content is dramatic and the coverage persists over a significant period of time.

Intensive news coverage, particularly in real time, diminishes the cushion of time that policy makers value highly. When people watch an event as it happens, they may expect a prompt response. Chinese tanks rolling into Tiananmen Square, a rocket attack on a Sarajevo market, refugees fleeing Kosovo—these and similar stories can generate particularly strong pressure on government decision-makers because not only do they capture the public's attention, they also elicit the public's sympathy. They want to see a response—now![17]

Policy makers, however, understand that complex political, military, and humanitarian layers usually lie beneath the events of the moment. Although a problem may be reported in real time, real-time solutions are hard to come by. The public, accustomed to high-speed headline versions of even the most complicated stories, may have little patience. This is when the CNN effect would come into play if policy makers were to succumb to public demands for action and respond too quickly. Policy making that becomes more reactive than proactive is almost certainly heading for problems.

This kind of CNN effect scenario has been superseded in recent years by a new version of media influence that might be called the "Al Jazeera effect." Indigenous media are cutting into Western media hegemony, supplanting the likes of CNN with local and regional news organizations. For policy makers, this has significant repercussions, as can be illustrated by comparing how people in the Middle East received information about two conflicts: the Gulf War of 1991 and the invasion of Iraq in 2003. In 1991, many people in Cairo, Amman, Beirut, and elsewhere in the Arab world most likely watched CNN or the BBC to get their war news because they had few options other than drab government-controlled stations. In 2003, they probably tuned in to Al Jazeera, Al Arabiya, or others of the new breed of Arab news stations. The key difference is not journalistic objectivity or technique, but credibility. CNN in Arabic was still CNN—an American-based company. Al Jazeera and the others are Arab-run and report the news from an Arab perspective, so regardless of esoteric debate about their journalism standards, they are more likely to be believed by their audience.

The impact of the "Al Jazeera effect" on Middle Eastern politics could first be seen in its reporting about the second Intifada, beginning in September 2000, which galvanized Arab support for the Palestinians. Some of the station's footage was picked up by networks around the world, perhaps most notably the coverage of the shooting of a twelve-year-old boy during a firefight between Palestinians and Israeli soldiers. Al Jazeera also critiqued Yasser Arafat's governance and described the conflict between the Palestinian Authority and Hamas. Hugh Miles wrote that as part of its Intifada coverage Al Jazeera also "routinely denounced the Egyptian government for failing the Arabs and there is no doubt these voices added to the destabilization in [Egypt]."[18]

Al Jazeera's status changed from regional to global player with its coverage from Afghanistan during the U.S. war against the Taliban beginning in 2001. Al Jazeera was allowed to remain in Taliban-controlled territory, and the station did not hesitate to show graphic pictures of civilian casualties, a practice it continued during the Iraq War. This graphic, real-time reporting undercut American efforts to portray the first conflict as a battle against terrorists and the second as a war of liberation, and it led some U.S. officials to accuse Al Jazeera of inflaming the "Arab street" and implicitly encouraging anti-American resistance.[19]

With so many Arab satellite channels now on the air (although not all are news-oriented), policy makers in the region and elsewhere must be prepared to deal with an expanding flow of real-time reporting that can increase the volatility of already intense situations. Governments also are unlikely to resist the temptation to try to manipulate real-time news coverage in ways that serve their purposes, which may range from straightforward articulation of policy to psychological warfare and deception. This is an important issue for news organizations, because, as William Prochnau has written, if they become caught up in the role of strategic enabler, their independence and credibility will be compromised.[20]

Real-time technology also demonstrates its political effectiveness at a more populist level in the form of blogging, which is open to anyone with access to the World Wide Web. Blogs are powerful amplifiers of voices that previously would have gone unheard, and as such they foster a degree of democratic parity at least in terms of expanding audience access for those who feel they have something worthwhile to say. The blogging firmament is already crowded and becoming more so, but bloggers are good at finding each other and reaching audiences. Particularly in countries where governments have tried to suppress political organizing, blogs may prove to be valuable tools in orchestrating pressure for reform.

In some instances bloggers serve a quasi-journalistic function—watching, critiquing, investigating. From the standpoint of policy makers and journalists, their speed and independence make them wild cards in various news business operations as well as in the political process. In 2004, CNN executive Eason Jordan made offhand comments at a conference in Switzerland to the effect that U.S. troops in Iraq had been shooting at journalists. Jordan later said he meant that the soldiers had been reckless, but bloggers reported the remarks as charges that the American military was targeting journalists. There was no transcript of Jordan's statement, and people who had been at the meeting offered various versions of what he had said, but the bloggers got the attention of mainstream news organizations. Their stories were no more definitive, but they generated so much heat that Jordan resigned. In 2005, bloggers supplied information that cautious (and often government-influenced) media did not provide as political change swept through Ukraine and Lebanon. The populist power of blog reporting can be seen in its ability to help bring people into the streets for political reasons.

Policy makers surveying the capabilities of real-time news in its many manifestations certainly realize that they must now utilize or compete with the din of many voices. The era of Western news organizations' hegemony is past, and the cushion of time that separated news reporting from policy responses has shrunk to wafer thinness. Like news, policy is now in the real-time world.

## CURRENT CHALLENGES IN PROVIDING COVERAGE

Despite technological advances that enhance real-time reporting, changes in the economic structure and operating principles of the news business make it impossible to say that the era of real-time news signals a true renaissance for international coverage by major news organizations. Stagnation in the mainstream news industry has been somewhat offset by the rise of issue- or event-specific Web sites, blogs, and other Internet-based reporting, but the mass news audience still relies on the big news organizations. The quality of their coverage has suffered, however, as their executives have made substantial cutbacks in overseas bureaus in important places such as Paris and Moscow, relying instead on "parachute journalism," which involves quickly sending a reporter to the spot where news is happening.

In terms of timeliness of reporting, this often works out because jet travel has shrunk the world to manageable size. But content may suffer significantly. Real-time reporting is not automatically good journalism; if coverage lacks substance, the speed of delivery means little. The parachute approach often produces news without context: the war or other humanitarian emergency appears to news consumers to have suddenly exploded, a distortion that occurs because journalists had not been on the scene to cover the situation while the fuse was burning. A reporter might arrive promptly and report breathlessly, but this kind of journalism is intrinsically misleading.

When covering complex international relations topics, episodic reporting—regardless of timeliness—is insufficient. As news organizations assess their approaches to international coverage in light of their expanded technological capacities, higher value should be placed on achieving consistency, even at the expense of speed.

On another front, the news media are likely to find increased competition from narrowly focused information providers that may have their own agendas, which they may seek to advance without regard to journalistic standards. The International Committee of the Red Cross, for example, might deliver its own real-time reports from a war zone to directly reach the ever-growing online audience. Another example: Hezbollah has its own television station, Al-Manar, which can reach a regional audience on the air and a global audience on the Internet.

These two very different organizations have (or can soon have) real-time reporting capability. Whether they would reach a sizable audience is problematic, but it may turn out that an increasingly Web-savvy global public will not be satisfied with traditional information sources and will rely more and more on specialized quasi-news sources such as these. Add to this mix the governments that use their own satellite channels or Internet venues to present their views directly to the public. There could soon be so many of these sources that even if each were to take just a sliver of audience away from mainstream news organizations, they could have significant cumulative economic effect (in terms of ratings/circulation-based revenues) and could make the broad—and often superficial—approach of "big news" less appealing to audience members who might know they could find information sources that speak directly to their specific interests. When these niche providers offer real-time coverage, they will be even more formidable competitors.

Making this more complicated is that on the Internet, information dissemination and political activity may share the same electronic terrain. Dorothy Denning writes: "The Internet can be an effective tool for activism, especially when it is combined with other communications media, including broadcast and print media and face-to-face meetings with policy makers. It can benefit individuals and small groups with few resources as well as organizations and coalitions that are large or well-funded. It facilitates such activities as educating the public and media, raising money, forming coalitions across geographical boundaries, distributing petitions and action alerts, and planning and coordinating events on a regional or international level."[21]

## TOWARD A NEW MODEL FOR REAL-TIME INTERNATIONAL NEWS COVERAGE

A critical task in improving international news coverage is determining how much reliance should be placed on real-time reporting and how much on other kinds of coverage. At the heart of this reform must be the recognition that faster is not necessarily better and that the news media have an obligation to educate rather than merely inform. This will require a commitment to presenting more substantive, context-defining stories even if they lack the showbiz flash of real-time reporting. This is not an all-or-nothing choice; real-time reports will remain an essential part of overall coverage but will not dominate.

Discretion in reliance on real-time news may help avoid situations such as occurred during the early days of the Iraq War when U.S. forces paused in their race toward Baghdad. Reporters on the scene and analysts in television studios wasted not a minute in flooding the public with commentary about the apparently stalled advance and flawed American war plan, retrieving from the buzzword attic the dreaded "quagmire," guaranteed to awaken unpleasant memories of Vietnam. They delivered their reports fast—real-time news in action—but they got it wrong. The pause was merely a pause; the invasion plan was brilliantly conceived and executed (despite the massive problems that followed).

This case underscores the flaws in the "faster is better" approach. A content analysis conducted by the Project for Excellence in Journalism found that during the first week of the war 60 percent of American television news stories were live and unedited and in 80 percent of the stories viewers heard from only the reporters, not from other sources.[22] Thomas Kunkel wrote that news organizations "showed an almost palpable need to rush to judgment. . . . How can you intelligently discuss military strategy four days into a war as complex as the invasion of Iraq? It seems to me this was another of those areas where the people, in their collective wisdom, were way ahead of the press. They implicitly understood that wars take time. While reporters were raising hell about tactics, every public survey demonstrated the American people were willing to let things play out before they made up their minds."[23]

This disconnect between news media and public is worth thinking about because it shows that in their quest to serve the public journalists may be proceeding down the wrong track. Tribune Publishing Company president Jack Fuller said the television coverage of the early stages of the war was "utterly riveting," but added that "it also demonstrated that there is a difference between seeing and understanding."[24] That distinction gets to the heart of the challenge facing journalism in the real-time era. Simply *seeing* can be accomplished with extraordinary breadth and precision; technology ensures that. But to help the public *understand* is far more difficult, and it will require the news media to occasionally take their feet off the accelerator as they provide information. To foster understanding will mean a new emphasis on careful, thoughtful construction of news stories in which a solid foundation of context supports the smaller structure of breaking news.

To appreciate these issues, consider this hypothetical case: You are an executive news producer at a major U.S. television network. You are receiving a real-time video feed of what appears to be a coup under way in an important Middle Eastern country. The video shows tanks surrounding the presidential palace, and your reporter on the scene—a stringer whom you have used only once or twice in the past—is telling you about rumors that the country's president has been killed.

So far, your competitors have put nothing on the air. What do you do? You might take the images and the stringer's report and broadcast them, adding a cautionary note about the unconfirmed nature of your coverage. You will be first, but will you be right? The alternative is to collect the material as it flows in, throw your resources into broader coverage, including corroboration and analysis about what these events mean, and then—when the story is pulled together—put it on the air. Even if other news organizations air the rumors and early footage without clear knowledge of what it means, caution on your part would be wise. Preliminary reporting should be accompanied with disclaimers about the incompleteness of the story. In taking this approach you might be undermining the notion of journalistic omniscience (which deserves undermining), and you might not be first with every scattered detail of the story, but you will be delivering coherent journalism, not merely acting as a conveyor of disconnected and questionable pieces of the story. Which approach would better serve the public?

In absolute terms, "faster is better" is true only when a story is of crucial, immediate importance to the audience (e.g., "The tornado is heading this way!"), and those stories are rare. No Luddite revolt against the technology of real-time coverage need occur, but that technology should simply aid editorial judgment, not drive it. If that principle is adopted, real-time news will still be valuable in its proper place in the journalistic hierarchy.

NOTES

1. Archibald MacLeish, "A Superstition Is Destroyed," encomium for dinner in honor of Edward R. Murrow, December 2, 1941, printed in *In Honor of a Man and an Ideal: Three Talks on Freedom* (n.p.: CBS, n.d.).

2. Jon Van, "Satellites Signal New Era in News Coverage, Viewing," *Chicago Tribune,* March 22, 2003; Michael Murrie, "New Technology Brings Live Coverage of the War in Iraq," *Communicator* (May 2003): 8.

3. Warren P. Strobel, *Late-Breaking Foreign Policy: The News Media's Influence on Peace Operations* (Washington, DC: U.S. Institute of Peace Press, 1997), 74.

4. Andrew Hoskins, *Televising War: From Vietnam to Iraq* (London: Continuum, 2004), 60.

5. Christopher Paul and James J. Kim, *Reporters on the Battlefield: The Embedded Press System in Historical Context* (Santa Monica, CA: RAND, 2004), 111.

6. Philip Seib, *Beyond the Front Lines: How the News Media Cover a World Shaped by War* (New York: Palgrave Macmillan, 2004), 60.

7. Philip Seib, *Going Live: Getting the News Right in a Real-Time, Online World* (Lanham, MD: Rowman and Littlefield, 2002), 142.

8. Michael Getler, "Yet Another Wake-Up Call," *Washington Post,* May 22, 2005.

9. Nielsen//Net Ratings, "American Web Surfers Boost Traffic to Foreign News Sites in March," Nielsen//NetRatings Web site, April 24, 2003, http://www.nielsen-netratings.com/pr/pr_030424.pdf.

10. Jacqueline E. Sharkey, "The Television War," *American Journalism Review* 25, no. 4 (May 2003): 20.

11. Bill Carter, "Nightly News Feels Pinch of 24-Hour News," *New York Times,* April 14, 2003.

12. Joseph J. Ellis, *His Excellency: George Washington* (New York: Knopf, 2004), 217; Kevin Phillips, *William McKinley* (New York: Henry Holt/Times Books, 2003), 94.

13. Michael J. Arlen, *Living-Room War* (1969; repr., New York: Penguin, 1982), 82.

14. Michael R. Beschloss and Strobe Talbott, *At the Highest Levels: The Inside Story of the End of the Cold War* (Boston: Little, Brown, 1993), 422, 431.

15. David Hoffman, "Global Communications Network Was Pivotal in Defeat of Junta," *Washington Post,* August 23, 1991; James A. Baker III with Thomas M. DeFrank, *The Politics of Diplomacy: Revolution, War and Peace, 1989–1992* (New York: Putnam's, 1995), 520; Philip Seib, *Headline Diplomacy: How News Coverage Affects Foreign Policy* (Westport, CT: Praeger, 1997), 112–14.

16. Stephen Badsey, "The Media and UN 'Peacekeeping' since the Gulf War," *Journal of Conflict Studies* 17, no. 1 (Spring 1997): 19.

17. David Perlmutter, *Photojournalism and Foreign Policy: Framing Icons of Outrage in International Crisis* (Westport, CT: Praeger, 1998).

18. Mohammed el-Nawawy and Adel Iskandar, *Al-Jazeera: The Story of the Network That Is Rattling Governments and Redefining Modern Journalism,* (2002; repr., Boulder, CO: Westview, 2003), 55–56; Hugh Miles, *Al-Jazeera: The Inside Story of the Arab News Channel That Is Challenging the West* (New York: Grove, 2005), 73, 71, 87.

19. Adel Iskandar and Mohammed el-Nawawy, "Al-Jazeera and War Coverage in Iraq," in *Reporting War: Journalism in Wartime,* ed. Stuart Allan and Barbie Zelizer (New York: Routledge, 2004), 325.

20. William Prochnau, "The Military and the Media," in *The Press,* ed. Geneva Overholser and Kathleen Hall Jamieson (New York: Oxford University Press, 2005), 325.

21. Dorothy Denning, "Activism, Hacktivism, and Cyberterrorism: The Internet as a Tool for Influencing Foreign Policy," in *Networks and Netwars: The Future of Terror, Crime, and Militancy,* ed. John Arquilla and David Ronfeldt (Santa Monica, CA: Rand, 2001), 242.

22. Project for Excellence in Journalism, "Embedded Reporters: What Are Americans Getting?" Project for Excellence in Journalism, April 3, 2003, www.journalism.org/resources/briefing/archive/war/embedding/default.

23. Thomas Kunkel, "Rushing to Judgment," *American Journalism Review* 25, no. 4 (May 2003): 4.

24. Sharkey, 20.

# 9

# AFTERWORD

## *Technology and the Policy Maker: No Place to Hide (or, Everyone Knows Everything)*

RICHARD MOOSE

### INTRODUCTION

The new information technology, platforms, and formats that have emerged within the past decade have transformed the relationship between policy makers and the world in which they function. This new relationship is due, in large measure, as the other contributors to this volume have demonstrated, to profound changes in how publics and policy makers receive and respond to foreign reporting. Underpinning most of what is new and compelling about this two-way flow is the Internet that links us all, both originators and users, within an ever-expanding sea of shared data.

Among the changes in reporting and response wrought by technology, none has been more transforming than the speed with which information now moves unless, perhaps, it is the exponential multiplication of producers. Virtually everyone who now receives via the new technology is also a potential originator. In addition, technology enables the aggregation of unprecedented volumes of information from widely diverse sources, and places it within easy and immediate reach of an ever-broader audience, including both old and new policy players. Each of these factors has made an identifiable impact on the way policy makers function.

Changes in the handling—and the handlers—of foreign news continue. The bombing of the World Trade Center and the subsequent controversy surrounding the policies and actions of the George W. Bush administration, as reported and played out in the news media, have intensified the interaction of policy makers and the media. At the height of the cold war, forty years ago, we did not experience the sustained domestic political tension that is evident today. We will examine the contrast between then and now as it relates to factors such as speed, access to information, new policy players, the American public, and, as a case study, look at how American policy makers are seeking to cope with the alarming erosion of respect for the United States abroad.

## THREE "WAR STORIES" OF SPEED AND IMMEDIACY

### The Cold War

More than forty years ago, as a junior foreign service officer, I returned to Washington from a post in West Africa where all of our telegrams came and went, handwritten and manually ciphered, through the local telegraph office. I was assigned to the State Department Operations Center, newly created in response to President Kennedy's complaint that he could never reach anyone at State after working hours. Along with a colleague, I passed the midnight hours in a windowless room with five telephones: one each to Number 10 Downing Street, the Elysée Palace, and the National Military Command Center (NMCC)—and two plain black civilian ones.

On several occasions, we were alerted by the NMCC that Soviet tanks had just rolled onto the German autobahn, closing Berlin's ground link with the Allied sectors of West Germany. We were comfortable, however, in the knowledge that six to eight hours might elapse before the rest of the world would know what we knew and that almost half a day would pass before government officials would be obliged to say anything to the press. The public was highly dependent in those days upon government-controlled sources—western or Soviet—for its information on such events. Moreover, the public generally got its news at most twice a day—once from the morning newspaper and again at the end of the day from the evening radio and television news broadcasts; there was no twenty-four/seven news bombardment like that we experience today.

If such an event could occur today, some Brandenburg farmer belatedly making his way home from a beer hall would see the tanks and flip a videophone image to his blogger pal, who would copy it to an Internet news site. The telephone of the press duty officer would be ringing before the farmer could get home and milk his cow. The more serious consequence, however, would be the chance that the commander of the U.S. Berlin Brigade would feel compelled to make a hurried judgment about how to respond. Put another way, fast-forwarding the decision process eliminates time for decompression and reflection.

### Vietnam

As U.S. troop levels in Vietnam were building in the 1960s and American press coverage of the war intensified, the Johnson administration became increasingly preoccupied with "handling the press." At the time, I was working for the national security adviser in the basement of the west wing of the White House. I began each day by listening over a poor telephone connection to the American public affairs briefing taking place twelve time zones away in Saigon. This daily event, referred to by journalist participants as the "Five O'Clock Follies," generally featured a recap of that day's events in Indochina from the official American perspective. It also provided an opportunity, eagerly seized upon by the journalists, to joust with U.S. press officials who, more often than not, had no firsthand knowledge of the matters they were presenting.

After listening to the give and take between briefers and journalists in Saigon, I would review the morning newspapers, scan the previous day's Vietnam speeches in the *Congressional Record* and convene a midmorning conference call from the Situation Room with selected public affairs offices around the government. Each would summarize what he proposed to say in response to the news of the day—actually the news of the day before—from the Far East. Often we would thrash out, on the telephone, a common administration line on the day's leading Indochina stories. Participants would then report back to their respective superiors, I to the National Security Council (NSC) adviser in the office adjoining mine and then, upstairs, to the president's press secretary. The State Department spokesman would check our proposed "line" with the secretary of state in time for the spokesman's regular eleven o'clock briefing. Much the same procedure was followed at the Pentagon, although the Defense Department generally tried to leave military comment to MACV (Military Assistance Command, Vietnam), its military headquarters in Saigon. The White House press office would normally brief journalists an hour after the State Department did, the delay affording us time to regroup should any of the earlier briefings have generated problems. Today's press handlers can only dream of such a deliberate pace.

The Johnson administration's credibility became a consuming issue by 1968 as firsthand reporting from Vietnam battered the president's insistence that we were winning the war. Most of this reporting originated from mainline American news organizations, supplemented by a few intrepid freelancers. In later years I came to know many of the journalists involved as a result of frequent visits to Indochina on behalf of the Senate Foreign Relations Committee, but as of January 1968, I was still in the White House basement working for the president's national security adviser.

There was no live video from Vietnam, and the filmed reporting that was flown out was normally at least twenty-four hours old before it was shown in the U.S. Nevertheless, television images of the war came to dominate the public's mind, none more so than those of January 31, 1968. On that day the North Vietnamese Army (NVA) and Vietcong (VC) launched a series of major attacks across South Vietnam in what became known as the Tet Offensive. When U.S. television networks aired film of VC fighters inside the walls of the U.S. Embassy compound in Saigon, the tide of American opinion turned irrevocably against American involvement.

In the early morning hours that same day, as the NSC adviser focused primarily on U.S. military reports of enormous VC casualties across the country, I drafted a cautiously defensive press statement. The president's adviser would have none of what I proposed but prepared his own draft, which, after some polishing in the press office, the president used at noon that day. The statement asserted that the NVA/VC attacks had come as no surprise and that our side would win. Coinciding with broadcast film of the fighting, all major U.S. papers featured grisly still photographs of the chief of the South Vietnamese National Police shooting a captured VC officer in the head with a pistol at close range. NBC ran color television film of the summary execution. Two months later, President Johnson announced that he would not run for reelection.

Vietnam remained a flaming public issue for several more years until President Nixon, after invading Cambodia and ending the draft, rapidly drew down U.S. troop levels. Only then did the intensity of public and journalistic pressure on the Nixon White House begin to subside, at least until, after the 1972 election, Watergate began to unfold.

The first American adviser in Vietnam was killed in July 1959. The first of our combat units—two battalions of marines—landed at Da Nang in March 1965. Saigon fell to the North Vietnamese in April 1975, after more than 47,000 American combat deaths. It is difficult to imagine another American armed engagement being sustained for over a decade—certainly not one fought like Vietnam then, or Iraq today. With the benefit of today's information environment, the public and the rest of the world would know too much to let that happen. The American public already knows far more today about political and religious factions in Iraq than it ever learned about internal Vietnamese political strife. Some journalists wrote about Vietnam in depth, and brilliantly, but they were not equipped to capture and transmit pictures of it with today's immediacy. No one took pictures of My Lai the morning after, whereas those who committed the atrocities at Abu Ghraib took pictures of themselves, pictures that the entire world would see. It is often said that Americans are impatient with war; perhaps it is the case that the more information they have, the less patient they become.

**Iraq**

We have seen how Vietnam coverage worked thirty years ago. Today, lightweight wireless equipment enables live coverage—security conditions permitting—from any part of Iraq at any hour. This capability has resulted in a succession of horrific images—live and otherwise—for those who are prepared to look. Bloggers, supplemented by a growing number of digital photographers, provide a diverse flow of information and commentary on every aspect of the war and U.S. occupation.

Where conditions are too dangerous for Westerners to go, Arab-language and other non-American media are often present to provide satellite feeds. Much of this coverage circulates daily throughout the Muslim world and beyond. U.S. viewers see little of it, although it is readily available to them on the Internet.

The Bush administration, like those of presidents Johnson and Nixon in previous years, today seeks to persuade the American public that the Iraq presented by the mainstream media is not reality. It is ironic that the media and the new technology—CNN reporters rolling live with the troops across the sands to Baghdad—having done so much to glamorize the U.S. military victory now have become the vehicle and photo chronicler of the occupation's mismanagement. Technology is not partisan, as my colleague David Perlmutter points out; the explosion is the story, and the media will show it regardless of whether it blows up the bad guys or the good guys.

## THE CONSEQUENCES OF SPEED AND OTHER LESSONS

The foregoing accounts illustrate how, over the last thirty to forty years, technology and new formats have reduced the distance—literal and figurative—between the public and events, often with telling political impact. Competitive pressure to be first on the scene with "breaking news" has further compounded the pressure on policy makers. As Steven Livingston writes in this volume, electronic audiences, now outnumbering those of print media, but with reduced attention spans and lacking the benefit of context, have been conditioned to expect reactions—and "breaking news"—now.

A veteran foreign affairs newscaster recently remarked to me, ruefully, following a blog-induced media flap, that instant communication puts "a premium on provocative comment rather than reasoned analysis." Similarly, a former senior policy official of the current administration asked about the role of analysis, complained, "React is the name of the game. Analysis? Sure, when there is time." This phenomenon works both ways, of course, on the policy maker as well, for two cycles are at work here, the news cycle driven by the new technology, and the policy cycle, increasingly driven by the news and the blogs, which have no cycle. They interact, the news cycle compressing the policy cycle and, with time, experience and analysis disappearing like matter into a black hole.

## DIVERSE SOURCES AND EXPANDED ACCESS

In addition to accelerating the movement of information, technology has vastly expanded the ranks of those who originate and disseminate information relevant to foreign policy. The Internet in particular—at once a resource as well as a broadcast and communications medium—has vastly broadened and facilitated access by the public to the substance of foreign policy. In the process, as we shall see, the policy maker has lost much of the exclusivity of function that he once enjoyed, acquiring at the same time both new competitors and new contenders for his attention.

Forty years ago, we aspiring policy makers believed that we knew it all and the public knew enough. As my colleague and I waited in the State Department Operations Center for President Kennedy to call (he never did), our other responsibility was to summarize ("short paragraphs and wide margins" was the guidance we were given) the most important diplomatic cables received overnight from U.S. missions abroad. Our product, known as "The Secretary of State's Morning Top Secret Summary," was not only read by the secretary but also delivered by courier to the White House, Defense Department, and perhaps to the CIA. Occasionally, our summary would include items from the wire services. This collection of information, plus the morning newspapers, framed the horizon of the secretary of state at the opening of business.

Our pride in the morning summary reflected the diplomat's subconscious belief that virtually everything of consequence in foreign policy was reflected in the flow of classified telegrams to and from diplomatic posts overseas. We shared—reluctantly—little of what we knew with those we judged had no "need to know," which included the press, Congress, and the American people.

Policy makers forty years ago generally were able to "frame" the public's view of foreign affairs. Journalists then depended, more often than not, upon government to tell them what was going on and what was important. Daily briefings at the Department of State were—but are no longer—a high point of the day for those who covered foreign affairs in Washington. The secrecy that so frequently cloaked government's intentions and actions during the cold war further constrained and channeled information flows. By and large the press accepted the need for secrecy, thus affording policy makers greater freedom of action—if only temporarily, as in the case of the Soviet tanks on the autobahn.

## TODAY, EVERYBODY KNOWS EVERYTHING

A goldfish bowl has replaced that windowless room where we once believed that we knew everything that was worth knowing. Diplomats and other officials, along with countless schools of other fishes, swim in that bowl, surrounded on all sides by an increasing number of twenty-four–

seven real-time commentators, Internet news Web sites, and an exponentially increasing number of bloggers.

Diplomats still seek to guard their private thoughts and agenda, but they repeatedly find themselves dealing in matters about which many other players are at least as well informed. Moreover, new players are continually distributing their own views directly to broad audiences and are demanding to know the diplomats'. A current official remarked recently, "We can't run and we can't hide. We used to be able to choose our issues. The choice is no longer ours."

## GOOD AS WELL AS BAD FOR POLICY MAKERS IN THE NEW TECHNOLOGY

There are pluses for the policy maker as well as minuses in what one of my colleagues has referred to as the new "information ecology." The new technology in journalism may have complicated the task of policy makers, but it has also enhanced the policy process in important ways. At working levels within the State Department, the access to information made possible by the Internet has expanded horizons and is breaking down the department's longstanding myopic fixation with its own "cable traffic."

The officer in charge of one highly sensitive country desk described to me the updated morning routine in the department. Everyone now begins with e-mail rather than "the cables," access to which once defined the hierarchy of the State Department. Now, because everyone knows that e-mail moves faster than the cables, everything of importance moves there first, flowing around and beyond the department's senior officialdom. Next, the desk officer will do a quick survey of leading news media, Internet news sites, and online newspapers, both domestic and foreign (no more reading the *Washington Post* at home or having to share the office copy of the *New York Times*). Finally, as time permits, come the cables from the field. Occasionally the officer surfs the Web sites of leading interest groups and relevant think tanks.

I have also been told that here and there in the Department of State there are those who read blogs. The senior official who confided this to me, however, hastened to add, "But it is mainly the young interns who do that." One hopes that they will not lose that habit as they age because, as Kaye Trammell and David Perlmutter write, blogs "will enrich the information the world receives about important events." Reaching out at the State Department like this for information would neither have been possible nor considered necessary—or desirable—even a decade ago.

## NEW PLAYERS, NEW ISSUES

Just as the State Department is now more disposed and able to look beyond its own product, outsiders now find it much easier to reach into the State Department process than ever before. This access is due largely to the leverage that the Internet has created. Many observers count this a good thing, for the State Department has often appeared protective of its overseas relationships at the expense of domestic constituencies. Attention to human rights is a case in point.

Thirty-five years ago, I served on the Foreign Relations Committee staff when the committee, urged on by Senator Clifford Case (R-NJ), initiated legislation to create the Human Rights Bureau at the State Department. A few years later, as assistant secretary of state for Africa, I encountered strong internal resistance from career officers as I sought to implement President Carter's activist human rights policy. Today, the press accords frequent attention to human rights violations, due in large measure to information developed and disseminated through the Internet by human rights–oriented nongovernmental organizations (NGOs). Policy makers can no longer afford to ignore NGOs, and the media coverage these groups command greatly strengthens the hand of those within the bureaucracy who advocate action to curb abuses.

Over the past decade diverse policy lobbying efforts have coalesced, thanks to the Internet. Some combinations, quite unlikely on the surface, have gained serious clout, such as Christian conservatives making common cause with liberal humanitarians, both seeking to end the atrocities at Darfur. A senior State Department negotiator attributes the now annual reports on "Trafficking in Persons" to "fierce lobbying by NGOs and successive administrations finally caving in to the incessant demands . . . not only by well-meaning souls but Christian and anti-abortion types who are nuts on the subject of prostitution and pornography." Depending on one's view of an issue, the enhanced access and prominence of public interest groups may enhance the policy making process.

On the other hand, the emergence of well-publicized new players and new issues has resulted in what one colleague refers to as a "multinodal foreign policy"—that is, diffuse policy centers. This development turns partly on personalities as well as the current administration's original antipathy to the State Department and partly on the emergence of a range of "multinational" issues. These include epidemic diseases, the environment, and, most potent of all, terrorism. Using e-mail, various groups inside the government promote and gather support for their points of view. These new, informal networks often bypass completely the State Department's traditional policy command and control mechanism, which in the past was based upon its monopoly on the creation, transmission, and dissemination of foreign policy cable traffic. Senior officers who have served recently in the State Department believe that the advent of e-mail has severely undermined the secretary of state's capacity to coordinate foreign policy.

## POLICY MAKERS MUST HAVE SPLIT SCREENS

Faced with a legion of new players, new issues, and constantly "breaking" stories, a policy maker today cannot afford to be caught far from cable TV news and Internet news sites. To ensure against surprise, most national security agencies now subscribe to commercial services that provide constantly updated electronic folders of instantly available worldwide coverage tailored to their needs. The policy maker now operates with a split screen: on one side what the public is being shown or told by the media; on the other, in-house information. It took the State Department longer than it should have to get there, but today, under the goad of an administration and State Department leadership that keys on public reactions, the State Department is learning to "go live with breaking news."

An American with senior experience in Moscow recounted to me an episode at the embassy there that helps to illustrate how the learning process worked. The year was 1992. The NSC staff at the White House was alarmed over a CNN report that a coup against President Yeltsin was under way in Moscow. An NSC staff person put in a telephone call to the embassy to ask why the embassy had not reported the coup attempt. As it happened, the embassy had not only reported but had even anticipated the disturbance. Its telegram had been received at the White House, but the staffer had not read it. On the Moscow end of the conversation, my friend listened while the chargé d'affaires (i.e., acting ambassador) patiently and repeatedly explained to the staffer that there was no cause for alarm. The latter insisted that a coup must be afoot because CNN said so. As the embassy had forecast, nothing came of the matter.

This probably was not the first such incident in American diplomatic history and certainly will not be the last. But coordination improved, as was evident later during the crisis between Boris Yeltsin and the Russian parliament. This time, it was the assistant to the president for national security affairs in Washington who watched on CNN as Russian forces surrounded the parliament building on orders from Yeltsin. Our ambassador to Moscow, trapped in the embassy by the shooting, dispatched a reporting team of officers still outside the perimeter into the streets, where they used a satellite telephone to share with Washington what they were learning on the ground. As the standoff wore on, the two policy makers agreed on a shared assessment of the situation and on the desirability of a gesture of support for Yeltsin. The signal was delivered not by diplomatic note but by a comment on CNN.

This arrangement worked to the policy makers' advantage on that occasion, but the NSC remained fearful of being caught on the wrong foot by CNN. The assistant to the president pressed the State Department to marshal major additional resources to beef up real-time, on-the-scene reporting capability in the future. As Under Secretary of State in overall charge of State Department resources at the time, I resisted, arguing that the State Department should not try to compete with CNN but instead analyze, monitor, correct, and supplement what was being broadcast, all of which was already within our capabilities—as had just been demonstrated in Moscow.

## ASSESSING THE PROCESS

It can be argued that American policy makers and the policy process should perform better today than forty years ago. Technology has enhanced policy makers' ability to gather information about what is going on in the world, to know what others are thinking (even without the

National Security Agency), to coordinate with one another instantaneously and to "put out their story." On the other hand, experience shows us that, in times of international crisis, facts, experts, and their analyses often are shoved aside as a greater number of decisions are pulled to the top, where, under any administration, political risk accumulates and political expediency is likely to trump expertise.

Like any set of tools, much depends on how the new information tools are used by policy makers. The existence and availability of superior technology for the collection and sharing of information proved to have little bearing on decision-making leading up to the invasion of Iraq or, thereafter, on the management of the occupation. The fog of war, the unique style of the Bush/Cheney administration, and the deep partisanship that it has engendered, make it difficult to render a judgment on the net benefit of enhanced information technology to current policy makers.

## AMERICA AND THE WORLD: A CASE STUDY

A wide range of observers across the ideological spectrum believe that America's negative image abroad is our most serious foreign policy challenge. The loss of respect that America has suffered around the world since 2003 goes far beyond the anti-Americanism that we knew in the past. Nor is it primarily the by-product of the self-abasing image of American culture exported by our own commercial media.

The intensity of reaction to American policies today can be related directly to advanced technologies wielded by non-U.S. actors over the Internet or in new news broadcasting operations. This intensity seems to have arisen from a fateful juxtaposition of events: the full flowering of new and unprecedented global news coverage and distribution capability just as the United States embarked upon a controversial policy of military action in a volatile region of the world already deeply mistrustful of U.S. motives.

New voices and new audiences like those in the Middle East have compounded the reaction to American policies. The most dramatic and troublesome example of this results from intensive coverage particularly by, but by no means limited to, Arabic-language media. This coverage, too often led by horrific images of our own creation, relentlessly stokes misunderstanding and mistrust of the United States. Research such as that reported in this volume by Margaret DeFleur confirms the geographic breadth and seriousness of the problem, particularly among the younger generations who will be their countries' leaders tomorrow.

## WILL "TRANSFORMATIONAL DIPLOMACY" BE THE ANSWER?

Observers agree on the severity of our loss of standing in the world, but they differ strongly over its cause. Conservatives are likely to blame terrorists and "radical Islam." Many others cite the style and content of Bush administration policies. Regardless of such differences, widespread support exists in political circles for an aggressive campaign of "public diplomacy."

Such a campaign was initiated in 2005 by the administration and was incorporated by Secretary Rice into the sweeping reallocation of State Department resources announced under the banner of "transformational diplomacy."[1] The unspoken premises of transformational diplomacy are that the administration's policies have not been forcefully defended by the State Department and thus are misunderstood; that critics at home and abroad have distorted our actions; and that more effective public relations, including a more rapid response to criticism, can redress the problem.

Beginning immediately and continuing over the next several years, hundreds of diplomats were to be shifted out of Washington and comfortable European posts to what Secretary Rice termed "critical emerging areas in Africa, South Asia, East Asia, the Middle East and elsewhere." But the aspect of the initiative that is of greatest interest in the context of our discussion is the heavy emphasis that transformational diplomacy places on use of electronic media. In the words of the department's press release, "Programs are being developed to enhance America's presence through a medium that young people worldwide increasingly rely upon for their information . . . programs to reach young people through interactive, online discussions." Regional media centers "will take America's story directly to the people and the regional television media in real time and in the appropriate language."

## PERSPECTIVES AND PROSPECTS FOR TRANSFORMATIONAL DIPLOMACY

Many things may be said about transformational diplomacy but, above all, these two: First, the announcement itself, while wrapped in the rhetoric of "promoting democracy" and "ending tyranny," constitutes an unmistakable acknowledgment of the scope and seriousness of the image problem that has emerged since the invasion and occupation of Iraq. Second, the very scale and nature of the diplomatic restructuring that is envisioned confirms, more clearly than any analysis, the enormous impact of information technology on the world of the policy maker. Here in her own words is Secretary Rice's assessment: "In the Middle East, for example . . . a vast majority of people get their news from a regional media network like Al Jazeera, not from a local newspaper. So our diplomats must tell America's story not just in translated op-eds, but live on TV in Arabic for a regional audience."

Secretary Rice has it right—up to a point. But it is essential to look beyond our current concerns over the Middle East and terrorism to appreciate the longer-term significance of indigenous satellite television. As Philip Seib points out, the more than 150 regional satellite television stations around the world are bringing an end to Western hegemony over world news broadcasting: the "Al Jazeera effect," Seib names it. This development is about far more than language; it is about masses of the heretofore uninformed who are seeing and hearing for the first time events that may affect their lives portrayed and interpreted in terms of their own culture rather than that of the Western industrialized Judeo-Christian democracies.

## EVEN IN IRAQ

I traveled to Baghdad in the late summer of 2003 on a futile mission to organize financial management training for officials of the new Iraqi government. One morning, instead of waiting to take an approved armored convoy, I hitched a ride into the city with an engineer whose company provided cell phone service to the U.S. Army. He described the burgeoning demand among Iraqis for cell phones and the resulting misuse of the occupation headquarters' cellular network. Along the way, the engineer pointed out groups of people on the sidewalks who he said were waiting to buy airtime minutes from "vendors" equipped with misappropriated U.S. government cell phones.

The point of the story is not about corruption (which is endemic) but to underscore how modern information technology continues to flood the world, even—or perhaps especially—war-torn countries like Iraq and Afghanistan, where I am told that the same is true. Before the Americans arrived, there may have been a few thousand tightly monitored cell phones in Iraq; in 2005 the research firm Strategy Analytics estimated the number at 3.5 million. Today, the number must be around 6 million, or one-third of the country's population. With every cell phone comes the opportunity to connect not just with family and friends but with the world at large through the Internet—now—with images, including of one's own neighborhood or city. In Steven Livingston's words, "one way or another, the globe is and will remain blanketed by devices that gather and distribute data, whether in the form of voice, video, digital stills, or text."

## A VERY DIFFERENT WORLD

Time will tell whether Secretary Rice's program can counter the Al Jazeera effect. If fully implemented, transformational diplomacy would represent the most extensive reallocation of resources within the State Department since the end of the cold war. Leaving aside the crucial factor of policy content, the other large question hanging over transformational diplomacy is whether it is adequate to the challenge posed by a global deluge of foreign affairs information. Secretary Rice's goal in remaking the State Department is anything but modest: she said, "America needs . . . a diplomacy that not only reports about the world as it is, but seeks to change the world itself."

## DÉJÀ VU

Forty-five years ago my wife, our two-month-old son, and I set out for a diplomatic assignment in West Africa. Ours was one of more than a dozen new U.S. embassies being opened with an eye to countering feared Soviet penetration of newly independent African nations. Ringing in our ears was John Kennedy's inaugural exhortation to "bear any burden, meet any hardship, support any friend, oppose any foe, in order to assure the survival and the success of liberty." One of our new friends and colleagues was the embassy public affairs officer, an engaging, talented, and energetic African American. It was his view that the greatest handicap of our information program in Africa was the legacy of American slavery.

In the 1960s and the decade following, the U.S. built and sustained an active diplomatic presence throughout Africa, including often-effective programs through the U.S. Information Agency (USIA). Technology did not figure prominently in these programs, either as a challenge or an asset. The shortwave Voice of America was as high tech as USIA got, and its penetration in Africa was minimal. There, as in most of the rest of the world in those days, the BBC was the standard of credibility. I recall that during my frequent visits to Indochina in the late 1960s and early 1970s, journalists and officials alike were listening the BBC, even my USIA friends. The latter explained to me that they tuned to the VOA to learn what they were expected to say but listened to BBC to find out what was going on.

Radio Moscow was hardly a threat to U.S. interests in Africa, where only a handful of elites listened to any foreign broadcasting at all. But our Soviet cold war adversaries blunted our efforts and achieved significant propaganda penetration without it. Their assets were mostly of our own making: racial violence in the United States, U.S. support for apartheid, and U.S. intervention in the Congo, Angola, and, later, Vietnam.

We are heading down a familiar road in seeking to counter Muslin extremism—looking for a technical fix to a political problem—i.e., that much of the rest of the world, whether or not it credits our good intentions, hates what the United States is doing in Iraq. Moreover, in this contest, the minuses of the revolution in information diffusion appear massively to outweigh any benefit to U.S. foreign policy operators. As we set out, in Secretary Rice's words, "to change the world itself," it would be good to remember that, at the end of the road, that which will matter most will be the actions behind the words and, above all, our credibility—not our technology.

NOTE

1. Condoleezza Rice, "Transformational Diplomacy," remarks at Georgetown University, January 16, 2006, posted at U.S. Department of State Web site, http://www.state.gov/rm/2006/59306.htm, accessed March 23, 2007.

# BIBLIOGRAPHY

"Airlines Find Help in Flights Overseas." *Wall Street Journal,* July 5, 2005.

Allbritton, Christopher. "Resume." Back to Iraq Weblog. http://www.back-to-iraq.com/personal/resume.html. Accessed September 11, 2005.

Allen, Cleo J. "Foreign News Coverage in Selected U.S. Newspapers, 1927–1997." Ph.D. diss., Louisiana State University, 2005.

Alterman, John B. "Transnational Media and Regionalism." *Transnational Broadcasting Studies* 1 (Fall 1998). http://www.tbsjournal.com/Archives/Fall98/Articles1/JA1/ja1.html.

Ambah, Faiiza Salah. "Suspects in Saudi Terror Plot Reportedly Include Teens." *Boston Globe,* June 23, 2003.

American-Arab Anti Discrimination Committee. "Protest Biased Media Coverage of Palestine and Palestinians." American-Arab Anti Discrimination Committee Web site, November 13, 2004. http://adc.org/index.php?id=2383.

American Society of Newspaper Editors. *Bringing the World Home: Showing Readers Their Global Connections.* American Society of Newspaper Editors, July 23, 1999, updated January 10, 2000. http://www.asne.org/index.cfm?ID=2569.

Anderson, James. "Questions of Democracy, Territoriality and Globalization." In *Transnational Democracy: Political Spaces and Border Crossings,* ed. James Anderson, 6–38. London: Routledge, 2002.

Arlen, Michael J. *Living-Room War.* 1969. Reprint, New York: Penguin, 1982.

Baard, Mark. "Reporter Takes His Weblog to War." *Wired,* March 14, 2003. http://www.wired.com/news/conflict/0,2100,58043,00.html.

Badsey, Stephen. "The Media and UN 'Peacekeeping' since the Gulf War." *Journal of Conflict Studies* 17, no. 1 (Spring 1997): 7–27.

Bailey, Charles W. "Foreign Policy and the Provincial Press." In *The Media and Foreign Policy,* ed. Simon Serfaty, 179–88. New York: St. Martin's, 1990.

Baker, James A., III, with Thomas M. DeFrank. *The Politics of Diplomacy: Revolution, War and Peace, 1989–1992.* New York: Putnam's, 1995.

Baker, Ray Stannard. "Marconi's Achievement." *McClure's,* February 1902, 291–99.

Bates, Stephen. *If No News, Send Rumors: Anecdotes of American Journals.* New York: Holt, 1989.

Baumgartner, Frank R., and Bryan Jones. *Agendas and Instability in American Politics.* Chicago: University of Chicago Press, 1993.

Benkler, Yochai. *The Wealth of Networks: How Social Production Transforms Markets and Freedom.* New Haven: Yale University Press, 2006.

Bennett, W. Lance. "The Burglar Alarm that just Keeps Ringing: A Response to Zaller." *Political Communication* 20, no. 2 (April–June 2003): 131–38.

Berger, Arthur Asa. *Cultural Criticism: A Primer of Key Concepts.* Thousand Oaks, CA: Sage Publications, 1995.

Beschloss, Michael R., and Strobe Talbott. *At the Highest Levels: The Inside Story of the End of the Cold War.* Boston: Little, Brown, 1993.

Bimber, Bruce A. *Information and American Democracy: Technology in the Evolution of Political Power.* New York: Cambridge University Press, 2003.

Boyd, Roderick. "Three Political Web Logs Make a Run for the Mainstream." *New York Sun,* May 3, 2005. http://www.nysun.com/article/13179.

Braudel, Fernand. *Civilization and Capitalism, 15th–18th Century.* Vol. 1, *The Structures of Everyday Life: The Limits of the Possible.* New York: Harper and Row, 1981.

Brown, Merrill. "Abandoning the News." *Carnegie Reporter* 3, no. 2 (Spring 2005): 3–8.

Brummitt, Chris. "U.S. Envoy 'Misspeaks' during Talk," *Baton Rouge Advocate,* October 22, 2005.

Burgess, John. "King Fahd Dies, Abdullah Succeeds." *Crossroads Arabia* Weblog, August 1, 2005. http://xrdarabia.org/blog/archives/2005/08/01/king-fahd-dies-abdullah-succeeds/.

Burkett, E., "Democracy Falls on Barren Ground." *New York Times,* March 29, 2005. http://www.nytimes.com/2005/03/29/opinion/29burkett.html.

Byrn, Edward W. "A Century of Progress in the United States." *Scientific American,* December 1900, 402–3.

Carnahan, Colleen. "KUVN-TV 23 Reigns as Top Rated Station in Dallas for Adults 18–34." Univisión Web site, June 20, 2005. http://www.univision.net/corp/en/pr/Dallas_20062005–1_print.html.

Carter, Bill. "Nightly News Feels Pinch of 24-Hour News." *New York Times,* April 14, 2003.

"Cell-Phone Sales to Reach 779 million This Year." MSNBC Web site, July 20, 2005. http://www.msnbc.msn.com/id/8641618/.

Cha, Ariana Eunjung. "Do-It-Yourself Journalism Spreads; Web Sites Let People Take News into Their Own Hands." *Washington Post,* July 17, 2005.

"Chinese Blogs Face Restrictions." BBC News Web site, June 7, 2005. http://news.bbc.co.uk/2/hi/technology/4617657.stm

Cobb, Roger W., and Charles D. Elder. *Participation in American Politics: The Dynamics of Agenda Building.* 2nd ed. Baltimore: Johns Hopkins University Press, 1983.

Connor, Alan. "Not Just Critics." *BBC News Magazine,* June 20, 2005. http://newswww.bbc.net.uk/1/hi/magazine/4111330.stm.

Cornfeld, Michael, et al. "Buzz, Blogs, and Beyond: The Internet and the National Discourse in the Fall of 2004." Pew Internet and American Life Project Web site, May 16, 2005. http://www.pewinternet.org/ppt/buzz_blogs_beyond_final05-16-05.pdf

Cox, Christopher. "Establishing Global Internet Freedom: Tear Down This Firewall." In *Who Rules the Net? Internet Governance and Jurisdiction,* ed. Adam Thierer and Clyde Wayne Crews Jr., 3–12. Washington, DC: Cato Institute, 2003.

De Moraes, Lisa. "Only CNN Gets the Picture." *Washington Post,* April 12, 2001.

Deedes, William Francis. *At War with Waugh: The Real Story of Scoop.* London: Macmillan, 2003.

———. "Evelyn Waugh in Ethiopia: Reflections and Recollections." *Journalism Studies* 2, no. 1 (February 2001): 27–29.

DeFleur, Melvin L., and Margaret H. DeFleur. *Learning to Hate Americans: How U.S. Media Shape Negative Attitudes among Teenagers in Twelve Countries.* Spokane: Marquette Books, 2003.

DeFleur, Melvin L., and Everette E. Dennis. *Understanding Mass Communication: A Liberal Arts Perspective.* Boston: Houghton Mifflin, 2002.

Denning, Dorothy. "Activism, Hacktivism, and Cyberterrorism: The Internet as a Tool for Influencing Foreign Policy." In *Networks and Netwars: The Future of Terror, Crime, and Militancy,* ed. John Arquilla and David Ronfeldt, 239–88. Santa Monica, CA: Rand, 2001.

Devenport, Mark. "Politicians Monitor the 'Bloggers.'" BBC News Web site, February 26, 2004. http://news.bbc.co.uk/1/hi/northern_ireland/3490568.stm.

Dimitrova, Daniela V., and Richard Beilock. "Where Freedom Matters: Internet Adoption among the Former Socialist Countries." *Gazette: The International Journal for Communication Studies* 67, no.2 (2005): 173–87.

Dotinga, Randy "A Blogger Shines When News Media Get It Wrong," *Christian Science Monitor,* August9, 2006.

Drezner, Daniel W., and Henry Farrell. "Web of Influence," *Foreign Policy,* November–December 2004, 32–41.

Dwight L. Morris and Associates. *America and the World: The Impact of September 11 on U.S. Coverage of International News.* Washington, DC: Pew International Journalism Program, 2002. Available on line at http://www.pewtrust.com/pdf/vf_pew_intl_fellows_911.pdf.

Ekins, H. R. *Around the World in Eighteen Days and How to Do It.* New York: Longmans, Green, 1936.

———. "A Reporter Aloft." In *We Cover the World: By Sixteen Foreign Correspondents,* ed. Eugene Lyons, 371–88. New York: Harcourt, Brace, 1937.

Elasmer, Michael G., and John E. Hunter. "The Impact of Foreign TV on a Domestic Audience: A Meta-analysis." *Communication Yearbook* 20 (1997): 47–69.

Ellis, Joseph J. *His Excellency: George Washington.* New York: Knopf, 2004.

Ellsworth, Brian. "Venezuela Launches Cable News Station." *NPR's Morning Edition,* National Public Radio, July 18, 2005. http://www.npr.org/templates/story/story.php?storyId=4758465.

El-Nawawy, Mohammed, and Adel Iskandar. *Al-Jazeera: The Story of the Network That Is Rattling Governments and Redefining Modern Journalism.* 2002. Reprint, Boulder, CO: Westview, 2003.

E.T. "About Families for San Diegan." View from Iran Weblog, May 27, 2005. http://viewfromiran.blogspot.com/2005/05/about-families-for-san-diegan-i-have.html.

Fenton, Tom. *Bad News: The Decline of Reporting, the Business of News, and the Danger to Us All.* New York: Regan Books, 2005.

Fialka, John J. *Hotel Warriors: Covering the Gulf War.* Washington, DC: Woodrow Wilson Center, 1992.

Fleeson, Lucinda. "Bureau of Missing Bureaus." *American Journalism Review* 25, no. 7, 2003: 32–39.

Flournoy, Don. "Coverage, Competition, and Credibility: The CNN International Standard." In *Global News: Perspectives on the Information Age,* 2nd ed., ed. Tony Silvia, 15–43. Ames: Iowa State University Press, 2001.

"Flier Sets Records in Tasman Sea Hop." *New York Times,* October 17, 1936.

Foreign Press Association of New York. "Writing about Everything That Is Worth Some News." Foreign Press Association of New York Web site. http://www.foreignpressnewyork.com/pages/members_reporting.htm. Accessed September 11, 2005.

Fradkin, Philip L. *Stagecoach: Wells Fargo and the American West.* New York: Simon and Schuster, 2002.

Frey, Linda S., and Marsha L. Frey. *The History of Diplomatic Immunity.* Columbus: Ohio State University Press, 1999.

Frost, Frank J. "The Dubious Origins of the 'Marathon.'" *American Journal of Ancient History* 4 (1979): 159–63.

Gandt, Robert L. *China Clipper: The Age of the Great Flying Boats.* Annapolis: Naval Institute Press, 1991.

Getler, Michael. "Yet Another Wake-Up Call." *Washington Post,* May 22, 2005.

Gillmor, Dan. *We the Media: Grassroots Journalism by the People, for the People.* Sebastopol, CA: O'Reilly Media, 2004.

Gitlin, Todd. "Public Sphere or Public Sphericules?" In *Media, Ritual, and Identity,* ed. Tamar Liebes and James Curran, 168–74. London: Routledge, 1998.

Glaser, Mark. "Nepalese Bloggers, Journalists Defy Media Clampdown by King." *Online Journalism Review,* February 23, 2005. http://www.ojr.org/ojr/stories/050223glaser/.

Glasser, Susan B. "Probing Galaxies of Data for Nuggets: FBIS Is Overhauled and Rolled Out to Mine the Web's Open-Source Information Lode." *Washington Post,* November 25, 2005.

Glines, Carroll V. *Round-the-World Flights.* New York: Van Nostrand Reinhold, 1982.

Grade, Michael. Prologue. *Future of the BBC.* BBC Web site, June 2004. http://www.bbc.co.uk/thefuture/bpv/prologue.shtml.

Guzzo, Glen. "Thinking Big: Covering Major International Stories Can Pay Significant Dividends for Regional Newspapers." *American Journalism Review* 26, no. 3 (June 2004): 20–22.

Hamilton, John Maxwell. *Main Street America and the Third World,* 2nd ed. Cabin John, MD.: Seven Locks Press, 1988.

——— and Eric Jenner. "Foreign Correspondence: Evolution, Not Extinction," *Nieman Reports* 58, no. 3 (Fall 2004): 98–100.

———. "Redefining Foreign Correspondence." *Journalism: Theory, Practice and Criticism* 5 (August 2004): 301–21.

Hannerz, Ulf. *Foreign News: Exploring the World of Foreign Correspondents.* Chicago: University of Chicago Press, 2004.

Harnett, Richard M., and Billy G. Ferguson. *Unipress: United Press International, Covering the 20th Century.* Golden, CO: Fulcrum, 2003.

Heingartner, Douglas. "Honoring News Photos as Picture-Taking Evolves." *New York Times,* May 3, 2005.

Hemming, John. *The Conquest of the Incas.* New York: Harcourt, Brace, Jovanovich, 1970.

Heppenheimer, T. A. *Turbulent Skies: A History of Commercial Aviation.* New York: Wiley, 1995.

Herodotus. *The Histories.* Trans. Aubrey de Selincourt. Rev. by A. R. Burn. New York: Penguin, 1972.

Hess, Stephen. *International News and Foreign Correspondents.* Washington, DC: Brookings Books, 1996.

Higgins, Jonathan. *Introduction to SNG and ENG Microwave.* Oxford, UK: Elsevier Focal Press, 2004.

Hoffman, David. "Global Communications Network Was Pivotal in Defeat of Junta." *Washington Post,* August 23, 1991.

Hoge, James F. "Foreign News: Who Gives a Damn?" *Columbia Journalism Review* 36, no. 4 (November–December 1997): 48–52.

Hohenberg, John. *Foreign Correspondence: The Great Reporters and Their Times.* New York: Columbia University Press, 1964.

———. "The New Foreign Correspondence." *Saturday Review,* January 11, 1969, 115–16.

Hoskins, Andrew. *Televising War: From Vietnam to Iraq.* London: Continuum, 2004.

Hudson, Frederic. *Journalism in the United States, from 1690 to 1872.* New York: Harper and Brothers, 1873.

Hunsaker, Jerome C. *Aeronautics.* Oxford, UK: Oxford University Press, 1952.

Iskandar, Adel, and Mohammed el-Nawawy. "Al-Jazeera and War Coverage in Iraq." In *Reporting War: Journalism in Wartime,* ed. Stuart Allan and Barbie Zelizer, 315–32. New York: Routledge, 2004.

Iyengar, Shanto. *Is Anyone Responsible? How Television Frames Political Issues.* Chicago: University of Chicago Press, 1991.

James, Meg. "Networks Have an Ear for Spanish." *Los Angeles Times,* September 11, 2005.

Kaplan, Robert D. "Get Me to Vukovar: The Lure of the Dangerous Road." *Columbia Journalism Review* 43 (September–October 2004): 11–12.

Karim, H. Karim. *From Ethnic Media to Global Media: Transnational Communication Networks among Diasporic Communities.* Hull, Quebec: Department of Canadian Heritage, 1998.

Kathmandu. "The Day Log." Radio Free Nepal Weblog, February 2, 2005. http://freenepal.blogspot.com/2005/02/day-log.html.

Keane, John. *Tom Paine, A Political Life.* 1995. Repr., New York: Grove, 2003.

Keck, Margaret E., and Kathryn Sikkink. *Activists beyond Borders: Advocacy Networks in International Politics.* Ithaca, NY: Cornell University Press, 1998.

Kent, Arthur. *Risk and Redemption: Surviving the Network News Wars.* Toronto: Viking, 1996.

Kingdon, John W. *Agendas, Alternatives, and Public Policies.* 1984. Reprint, New York: HarperCollinsCollege, 1995.

Knowlton, Brian. "Bush Faces Rising Complaints about Handling of Disaster." *New York Times,* September 4, 2005.

Kristof, Nicholas D. "Terrorists in Cyberspace." *New York Times,* December 20, 2005.

Kroeger, Brooke. *Nellie Bly: Daredevil, Reporter, Feminist.* New York: Times Books, 1994.

Kulikova, Svetlana V., and David D. Perlmutter. "Blogging Down the Dictator? The Kyrgyz Revolution and Samizdat Websites." *International Communication Gazette* 69, no. 1 (2007): 29–50.

Kunkel, Thomas. "Rushing to Judgment." *American Journalism Review* 25, no. 4 (May 2003): 4.

Kurtz, Howard. "In the Blogosphere, Lightning Strikes Thrice." *Washington Post,* February 13, 2003.

Laughland, John. "The Mythology of People Power." *Guardian,* April 1, 2005.

Lawrence, Regina. *The Politics of Force: Media and the Construction of Police Brutality.* Berkeley: University of California Press, 2000.

Leonardi, Paul M. "Problematizing 'New Media': Culturally Based Perceptions of Cell Phones, Computers, and the Internet among United States Latinos." *Critical Studies in Media Communication* 20, no. 2 (June 2003): 160–79.

Leontiev, M. "Kirgizskiy zvonok dlya Rossii" [Kyrgyzstan Rings the Bell for Russia]. *Komsomolskaya Pravda,* March 28, 2005. Translation provided by Svetlana Kulikova.

Lewis, Sian. *News and Society in the Greek Polis.* Chapel Hill: University of North Carolina Press, 1996.

Livingston, Steven. "Diplomacy and Remote-Sensing Technology: Changing the Nature of Debate." *Net Diplomacy: 2015 and Beyond,* no.16 (2002): 1–7.

———. "Media Coverage of the War: An Empirical Assessment." In *Kosovo and the Challenge of Humanitarian Intervention,* ed. Albrecht Schnabel and Ramesh Thakur, 360–84. Tokyo: United Nations University Press, 2001.

———. "The New Information Environment and Diplomacy." In *Cyber-Diplomacy in the Twenty-first Century,* ed. Evan H. Potter, 110–27. Montreal: McGill-Queen's University Press, 2002.

———. "Remote Sensing Technology and the News Media." In *Commercial Observation Satellites: At the Leading Edge of Global Transparency*, ed. John Baker, Kevin O'Connell, and Ray Williamson, 485–502. Santa Monica, CA: Rand Corporation and the American Society for Photogrammetry and Remote Sensing, 2001.

———. "Transparency and the News Media." In *Power and Conflict in the Age of Transparency*, ed. Bernard I. Finel and Kristin M. Lord, 257–85. New York: Palgrave, 2000.

Livingston, Steven, and W. Lance Bennett, "Gatekeeping, Indexing and Live-Event News: Is Technology Altering the Construction of News?" *Political Communication* 20, no. 4 (October–December 2003): 363–80.

Livingston, Steven, and Todd Eachus. "Humanitarian Crises and U.S. Foreign Policy: Somalia and the CNN Effect Reconsidered." *Political Communication* 12, no. 4 (October–December 1995–96): 413–29.

Livingston, Steven, and Douglas Van Belle. "The Effects of New Satellite Newsgathering Technology on Newsgathering from Remote Locations." *Political Communication* 22, no.1 (January–March 2005): 45–62.

Logsdon, Tom. *Mobile Communication Satellites.* New York: McGraw-Hill, 1995.

Lubow, Arthur. *The Reporter Who Would Be King: A Biography of Richard Harding Davis.* New York: Scribner, 1992.

MacLeish, Archibald. "A Superstition Is Destroyed." Encomium for dinner in honor of Edward R. Murrow, December 2, 1941. In *In Honor of a Man and an Ideal: Three Talks on Freedom* (n.p.: CBS, n.d.).

Mallery, Garrick. *Sign Language among North American Indians, Compared with That among Other Peoples and Deaf Mutes.* 1881. Reprint, The Hague: Mouton, 1972.

Marks, Jason. *Around the World in 72 Days: The Race between Pulitzer's Nellie Bly and* Cosmopolitan*'s Elizabeth Bisland.* New York: Gemittarius, 1993.

Marvin, Carolyn. *When Old Technologies Were New: Thinking about Electric Communication in the Late Nineteenth Century.* New York: Oxford University Press, 1988.

Maass, Peter. "Salam Pax Is Real." *Slate,* June 2, 2003. http://slate.msn.com/id/2083847/.

Miles, Hugh. *Al-Jazeera: The Inside Story of the Arab News Channel That Is Challenging the West.* New York: Grove, 2005.

Mindich, David T. Z. *Just the Facts: How "Objectivity" Came to Define American Journalism.* New York: New York University Press, 1998.

Miranda, Frankie. "Univision News #1 in Gotham, 4th Time in Two Weeks," Univisión Web site, August 4, 2005. http://www.univision.net/corp/en/pr/New_York_04082005-2_print.html.

Mobiledia, "Camera Phones to Steal Low-End Digital Camera Market." Mobiledia Web site, August 11, 2005. http://www.mobiledia.com/news/34302.html.

Modley, Rudolf. *Aviation Facts and Figures, 1945*. New York: McGraw-Hill, 1945.

Morganstein, Brooke, Cassandra Bujarski, and Shannon Provost. "Univision Shakes Up ABC, CBS, NBC and FOX in Historic 2004–2005 Season." Univisión Web site, May 2, 2005. http://www.univision.net/corp/en/pr/Los_Angeles_02052005-2.html.

Motlagh, Jason. "Words Are Weapons for Iranian Bloggers." United Press International, February 17, 2005. http://washtimes.com/upi-breaking/20050214-050322-8970r.htm.

Mowrer, Paul Scott. *House of Europe*. Boston: Houghton Mifflin, 1945.

Murrie, Michael. "New Technology Brings Live Coverage of the War in Iraq." *Communicator*, May 2003, 6–8.

New California Media. "First-Ever Quantitative Study on the Reach, Impact and Potential of Ethnic Media." New America Media Web site, April 22, 2002. http://news.ncmonline.com/news/view_article.html?article_id=796.

"A New Front in Phone Fight." *International Herald Tribune*, August 27–28, 2005.

Nielsen//NetRatings. "American Web Surfers Boost Traffic to Foreign News Sites in March." Nielsen//NetRatings Web site, April 24, 2003. http://www.nielsen-netratings.com/pr/pr_030424.pdf.

Noble, Iris *Nellie Bly, First Woman Reporter*. New York: Messner, 1956.

Northey, Sue. *The American Indian*. 1939. Reprint, San Antonio: Naylor, 1954.

Owens, William A., with Edward Offley. *Lifting the Fog of War*. New York: Farrar, Straus and Giroux, 2000.

Parks, Michael. "Foreign News: What's Next?" *Columbia Journalism Review* 40, no. 5 (2002): 52–57.

Paul, Christopher, and James J. Kim. *Reporters on the Battlefield: The Embedded Press System in Historical Context*. Santa Monica, CA: RAND, 2004.

Perlmutter, David D. "Bird Flu: Blogging Truth to Power." PolicyByBlog Weblog, January 9, 2006. http://policybyblog.squarespace.com/journal/2006/1/7/bird-flu-blogging-truth-to-power.html.

———. *Blogwars: The New Political Battleground*. New York: Oxford University Press, forthcoming.

———. *Photojournalism and Foreign Policy: Framing Icons of Outrage in International Crisis.* Westport, CT: Praeger, 1998.

———. "Will Blogs Go Bust?" *Editor and Publisher,* August 4, 2005. http://www.editorandpublisher.com/eandp/article_brief/eandp/1/1001009362.

Perrone, Jane. "Attack on London, 11:18 a.m.," *Guardian* Web site, July 22, 2005. http://blogs.guardian.co.uk/news/archives/2005/07/22/she_was_moaning_out_of_pure_horror_and_terror.html

Phillips, Kevin. *William McKinley.* New York: Henry Holt/Times Books, 2003.

Pillersdorf, Stephanie, and Brooke Morganstein. "Univision Announces 2005 Second Quarter Results." Univisión Web site, August 4, 2005. http://www.Univision.net/corp/en/ir/2q05.pdf.

Posner, Richard A. "Bad News." *New York Times Book Review,* July 31, 2005, 1–14.

Powers, William. "Hello, World." *National Journal* 33, no. 26 (2001): 2082.

Prochnau, William. "The Military and the Media." In *The Press,* ed. Geneva Overholser and Kathleen Hall Jamieson, 310–31. New York: Oxford University Press, 2005.

Project for Excellence in Journalism, "Embedded Reporters: What Are Americans Getting?" Project for Excellence in Journalism, April 3, 2003. www.journalism.org/resources/briefing/archive/war/embedding/default.

Rainie, Lee. "The State of Blogging." Pew Internet and American Life Project Web site, January 2005. http://www.pewinternet.org/pdfs/PIP_blogging_data.pdf.

Ramos, Jorge. *No Borders: A Journalist's Search for Home.* Trans. Patricia J. Duncan. New York: Rayo, 2002.

Rice, Condoleezza. "Transformational Diplomacy." Remarks at Georgetown University, January 16, 2006. U.S. Department of State Web site, http://www.state.gov/rm/2006/59306.htm. Accessed March 23, 2007.

Rodríguez, América. "Creating an Audience and Remapping a Nation: A Brief History of U.S. Spanish-Language Broadcasting, 1930–1980." *Quarterly Review of Film and Video* 16, no. 3–4 (1999): 357–74.

———. "Objectivity and Ethnicity in the Production of the Noticiero Univision." *Critical Studies in Mass Communication* 13 (1996): 59–81.

Rojas, Viviana. "The Gender of Latinidad: Latinas Speak about Hispanic Television." *Communication Review* 7, no. 2 (April–June 2004): 125–53.

Rosenblum, Mort. *Coups and Earthquakes: Reporting the World for America.* New York: Harper and Row, 1979.

Rosenthal, Elisabeth. "The Cellphone as Church Chronicle: Creating Digital Relics." *New York Times,* April 8, 2005.

Salhani, Claude. "Politics and Policies: The Other Mideast Revolt." United Press International, March 2, 2005.

Saltzman, Jonathan. "Far from the Front, Cases of Anxiety Rise." *Boston Globe,* April 6, 2003.

"Salam's Story." *Guardian* Web site, May 30, 2003. http://www.guardian.co.uk/Iraq/Story/0,2763,966819,00.html.

Savoor, Namrata. "'Parachute Journalism' Hurts World News Overseas." Freedom Forum Web site, May 30, 2001. http://www.freedomforum.org/templates/document.asp?documentID=14034.

Schudson, Michael. *Discovering the News: A Social History of American Newspapers.* New York: Basic Books, 1978.

Schwartz, John. "Blogs Provide Raw Details from Scene of Disaster" *New York Times,* December 28, 2004.

Schramm, Wilbur, Jack Lyle, and Edwin Parker. *Television in the Lives of Our Children.* Stanford, CA: Stanford University Press, 1961.

Seelye, Katharine Q. "Newsweek Apologizes for Report of Koran Insult." *New York Times,* May 16, 2005.

Seib, Philip. *Beyond the Front Lines: How the News Media Cover a World Shaped by War.* New York: Palgrave Macmillan, 2004.

———. *Going Live: Getting the News Right in a Real-Time, Online World.* Lanham, MD: Rowman and Littlefield, 2002.

———. *Headline Diplomacy: How News Coverage Affects Foreign Policy.* Westport, CT: Praeger, 1997.

Seplow, Stephen. "G.A.s for the World." *American Journalism Review* 25, no. 7 (October–November 2003): 40–45.

Shanor, Donald R. *News from Abroad.* New York: Columbia University Press, 2003.

Sharkey, Jacqueline E. "The Television War." *American Journalism Review* 25, no. 4 (May 2003): 18–27.

Shaw, David. "Foreign Correspondents: It's On-the-Job Training." *Los Angeles Times,* July 2, 1986.

———. "Foreign News Shrinks in Era of Globalization." *Los Angeles Times,* September 27, 2001.

Sigal, Leon V. *Reporters and Officials: The Organization and Politics of Newsmaking.* Lexington, MA: Heath, 1973.

Smith, C. S. "Kyrgyzstan's Shining Hour Ticks Away and Turns Out to Be a Plain, Old Coup." *New York Times,* April 3, 2005.

Smith, Richard Norton. *The Colonel: The Life and Legend of Robert R. McCormick, 1880–1955.* Boston: Houghton Mifflin, 1997.

Smith, Robert C. "Comparing Local-Level Swedish and Mexican Transnational Life: An Essay in Historical Retrieval." In *New Transnational Social Spaces: International Migration and Transnational Companies in the Early Twenty-first Century,* ed. Ludger Pries, 37–58. New York: Routledge, 2001.

Snyder, Alvin. "Journalism: A Risky Profession." USC Center on Public Diplomacy Web site, July 21, 2005. http://www.ebu.ch/CMSimages/en/USC%20Center%20on%20Public%20Diplomacy%20_%20HEST_tcm6-43172.pdf.

Southern, P. *Signals versus Illumination on Roman Frontiers.* London: Britannia, 1990.

Spencer, R. "American Helped Plant Tulip Uprising." *Ottawa Citizen,* April 2, 2005.

Spychalski, John C. "Transportation." In *The Columbia History of the 20th Century,* ed. Richard W. Bulliet, 403–36. New York: Columbia University Press, 1998.

Standage, Tom. *The Victorian Internet: The Remarkable Story of the Telegraph and the Nineteenth Century's On-Line Pioneers.* New York: Berkley Books, 1999.

Starr, Paul. "The Meaning of Abu Ghraib." *American Prospect,* June 2004. http://www.princeton.edu/~starr/articles/articles04/Starr-MeaningAbuGhraib-6-04.htm.

Stoppard, Tom. *Night and Day.* New York: Samuel French, 1978.

Strobel, Warren P. *Late-Breaking Foreign Policy: The News Media's Influence on Peace Operations.* Washington, DC: U.S. Institute of Peace Press, 1997.

Strupp, Joe. "Nouveau Foreign Correspondents' Tastes Range from Wine to War." World Press Institute Web site, 2000. http://www.worldpressinstitute.org/internat.htm.

Su, Norman M. "A Bosom Buddy Afar Brings a Distant Land Near: Are Bloggers a Global Community?" Paper presented at the Second International Conference on Communities and Technologies (C&T 2005).

Suárez-Orozco, Marcelo M., and Mariela M. Páez. "Introduction: The Research Agenda." In *Latinos: Remaking America,* ed. Marcelo M. Suárez-Orozco and Mariela M. Páez, 1–137. Berkeley: University of California Press, 2002.

Subervi, Federico. *Network Brownout 2004: The Portrayal of Latinos and Latino Issues in Network Television News.* Austin and Washington, DC: National Association of Hispanic Journalists, 2004.

Sullivan, E. "A Scary Democratic Rebellion in Kyrgyzstan." *Cleveland Plain Dealer,* March 27, 2005.

Suro, Roberto. "Changing Channels and Crisscrossing Cultures: A Survey of Latinos and the News Media." Pew Hispanic Center, April 2004. http://pewhispanic.org/files/reports/27.pdf.

Tait, Richard. "The Future of International News on Television." *Historical Journal of Film, Radio and Television* 20, no. 1 (2000): 51–53.

Technorati. "About Us." Technorati Web site, 2005. http://www.technorati.com/about/.

Thomas, Rosalind. *Oral Tradition and Written Record in Classical Athens.* Cambridge: Cambridge University Press, 1989.

Trammell, Kaye D. "Celebrity Blogs: Investigation in the Persuasive Nature of Two-Way Communication Regarding Politics." Ph.D. diss., University of Florida, 2004.

———. "Looking at the Pieces to Understand the Whole: An Analysis of Blog Posts, Comments, and Trackbacks." Paper presented at annual meeting of International Communication Association, New York, 2005.

Turow, Joseph. *Media Today: An Introduction to Mass Communication.* Boston: Houghton Mifflin, 1999.

UN Department of Economic and Social Affairs. "World Economic and Social Survey: International Migration." United Nations Web site, 2004. http://www.un.org/esa/policy/wess/wess2004files/part2web/preface.pdf.

———, Population Division. "World Population by Gender." New York: United Nations, 2003.

"Univision Media Properties: Univision and Telefutura Television Groups," Univisión Web site. http://www.univision.net/corp/en/utg.jsp.

Van, Jon. "Satellites Signal New Era in News Coverage, Viewing." *Chicago Tribune,* March 22, 2003.

Varadarajan, Tunku. "Parachute Journalism Redux." *Wall Street Journal,* November 12, 2001.

Vertovec, Steven. "Conceiving and Researching Transnationalism." *Ethnic and Racial Studies* 22, no. 2 (1999): 447–62.

Waugh, Evelyn. *Scoop.* Boston: Little, Brown, 1937.

Wendland, Mike. "From ENG to SNG: TV Technology for Covering the Conflict with Iraq." Poynteronline Weblog, March 6, 2003. http://www.poynter.org/content/content_view.asp?id=23585.

Wiley, Stephen B. Crofts. "Rethinking Nationality in the Context of Globalization." *Communication Theory* 14, no. 1 (2004): 78–96.

Wilgenburg, Vladimir van. "Turkish State Not Helping Kurds Dying from Flue." Weblog entry, http://vladimirkurdistan.blogspot.com/2006/01/turkish-state-not-helping-kurds-dying.html.

Williams, Andrew Paul. "Media Narcissism and Self-Reflexive Reporting: Metacommunication in Televised News Broadcasts and Web Coverage of Operation Iraqi Freedom." Ph.D. diss., University of Florida, 2004.

Wilson, G. Lloyd, and Leslie A. Bryan. *Air Transportation.* New York: Prentice-Hall, 1949.

Wilson, Geoffrey. *The Old Telegraphs.* London: Phillimore, 1976.

Wood, Nicholas. "Video of Serbs in Srebrenica Massacre Leads to Arrest." *New York Times,* June 3, 2005.

"The World in Figures," 2004. In *The World in 2003,* special issue of *The Economist,* November 2002, 81–87.

World Resources Institute. "Civil Society: International Non-Governmental Organizations with Membership" (table). World Resources Institute Earth Trends online database, 2005. http://earthtrends.wri.org/text/environmental-governance/variable-575.html.

World Travel and Tourism Council. *World Travel and Tourism: Sowing the Seeds of Growth.* World Travel and Tourism Council Report no. 6. London: World Travel and Tourism Council, 2005.

Xiao Qiang. "Chinese Whispers." *New Scientist,* November 27, 2004. http://technology.newscientist.com/channel/tech/mg18424755.500-chinese-whispers-.html.

Yuferovam, Y. "Kirgizskiy perevorot" [The Kyrgyz Coup]. *Rossiyskaya Gazeta,* March 30, 2005. Translation provided by Svetlana Kulikova.

Zittrain, Jonathan, and Benjamin Edelman. "Empirical Analysis of Internet Filtering in China." Berkman Center for Internet and Society, Harvard Law School, 2003. http://cyber.law.harvard.edu/filtering/china/.

# CONTRIBUTORS

MARGARET H. DEFLEUR is the Associate Dean of Graduate Studies and Research at the Manship School of Mass Communication at Louisiana State University. Previously, DeFleur was director of the master's program in health communication at Boston University. Earlier academic positions include visiting scholar at Harvard University's School of Public Health and adjunct professor at the S. I. Newhouse School of Public Communications at Syracuse University. Her teaching and research interests include media theory and effects, health communication, and new media. Her recent works include *Fundamentals of Human Communication: Social Science in Everyday Life* (McGraw-Hill, 2005), *Learning to Hate Americans: How. U.S. Media Shape Attitudes among Teenagers in Twelve Nations* (Marquette Books, 2003, coauthored with Melvin L. DeFleur), and articles published in academic journals.

EMILY ERICKSON is an assistant professor at Louisiana State University's Manship School of Mass Communication. Although she has written conference papers, articles, and book chapters on topics ranging from agenda setting to campaign finance, Dr. Erickson specializes in First Amendment jurisprudence and is particularly interested in the role of corporations in the modern marketplace of ideas. She is the coeditor of *Contemporary Media Issues,* 2nd edition (Vision Press, 2004).

JOHN MAXWELL HAMILTON came to Louisiana State University in 1992 after more than two decades as a journalist and public servant. Hamilton reported abroad for ABC Radio and the *Christian Science Monitor,* among other media, and was a longtime national commentator on Public Radio

International's *Marketplace.* Hamilton served in the U.S. Agency for International Development during the Carter administration, on the staff of the House Foreign Affairs Committee, and at the World Bank. He was the first to explore systematic ways to improve local coverage of foreign affairs and has played a leading role in shaping public opinion about U.S.–Third World relations. He is a member of the Council on Foreign Relations and on the board of the International Center for Journalists. Hamilton is author or coauthor of five books and has been a Pulitzer Prize juror and a fellow of the Joan Shorenstein Center on the Press, Politics and Public Policy at Harvard University. He was appointed the Louisiana State University Foundation Hopkins P. Breazeale Professor in 1998. The Freedom Forum named him the 2003 Journalism Administrator of the Year.

STEVEN LIVINGSTON is a professor of political communication in the School of Media and Public Affairs at George Washington University, where he holds a joint appointment in the Elliott School of International Affairs. He is also a research professor in the Political Science Department and is a faculty associate in GWU's Space Policy Institute. Livingston's research and teaching focus on media, advanced information technology, and international affairs. He is also chairman of the board of the Public Diplomacy Institute, an organization within SMPA that he cofounded. He has lectured at the Naval War College, the Army War College, and the National Defense University and has spoken at the Carnegie Endowment for International Peace, the U.S. Institute of Peace, and the Rand Corporation, among other institutions. His most recent book is *When the Press Fails: Political Power and the News Media from Iraq to Katrina,* with W. Lance Bennett and Regina Lawrence (University of Chicago Press, 2007).

RICHARD MOOSE was once a career foreign service officer, later Assistant Secretary of State for African Affairs, and then Under Secretary of State for Management. He was also a senior staff member of the Committee on Foreign Relations of the U.S. Senate, special assistant to the National Security Adviser, and staff secretary of the National Security Council. Between tours in the government, Moose was a managing director of Shearson Lehman Brothers and Senior Vice President of the American Express Company. Today he is active in civic affairs in Alexandria, Virginia.

LISA PAULIN is an assistant professor of Mass Communication at North Carolina Central University. Her research focuses on news representation of Latinos in emerging immigrant communities and Latino-oriented media. Paulin received her bachelor's and master's degrees from Indiana University and her Ph.D. from the University of North Carolina at Chapel Hill. She taught at the Monterrey Institute of Technology and Higher Studies in Mexico City, Mexico, prior to beginning her Ph.D.

DAVID D. PERLMUTTER was, at the time this book was written, an associate professor at the Manship School of Mass Communication at Louisiana State University and a senior fellow at the Reilly Center for Media and Public Affairs. In July 2006 he became a professor and associate dean for graduate studies and research at the William Allen White School of Journalism and Mass Communications, University of Kansas. With a B.A. and M.A. from the University of Pennsylvania, he received his Ph.D. from the University of Minnesota. A documentary photographer, he is the author of *Photojournalism and Foreign Policy: Framing Icons of Outrage in International Crises* (Praeger, 1998), *Visions of War: Picturing Warfare from the Stone Age to the Cyber Age* (St. Martin's, 1999), *Policing the Media: Street Cops and Public Perceptions of Law Enforcement* (Sage, 2000), and *Picturing China in the American Press: The Visual Portrayal of Sino-American Relations in* Time *Magazine, 1949–1973* (Rowman and Littlefield, 2007) and editor of *The Manship School Guide to Political Communication* (Louisiana State University Press, 1999). His *Blogwars: The New American Political Battleground* is forthcoming from Oxford University Press. He has also written over 150 opinion essays for U.S. and international newspapers and is a columnist for the *Chronicle of Higher Education.* He is the editor of the blog of the Robert J. Dole Institute of Politics (http://www.doleinstitute.org).

PHILIP SEIB is Professor of Journalism and Public Diplomacy at the Annenberg School for Communication, University of Southern California. Previous to that he was Lucius W. Nieman Professor of Journalism at Marquette University. Seib is the author of numerous books, including *Headline Diplomacy: How News Coverage Affects Foreign Policy* (Praeger, 1997), *Going Live: Getting the News Right in a Real-Time, Online World* (Rowman and Littlefield, 2001), *The Global Journalist: News and Conscience in a World of Conflict* (Rowman and Littlefield, 2002), and, most recently, *Beyond the Front Lines: How the News Media Cover a World Shaped by War* (Palgrave Macmillan, 2004). Seib worked for many years as a print and television journalist, and his newspaper columns and television reporting about politics and social issues won numerous awards. He holds degrees from Princeton University and Southern Methodist University.

KAYE SWEETSER TRAMMELL was, at the time this book was written, an assistant professor at the Manship School of Mass Communication at Louisiana State University and cowrote her contribution with the support of a grant from the Reilly Center for Media and Public Affairs. She is currently an assistant professor of public relations in the Grady College of Journalism and Mass Communication at the University of Georgia. Her research focuses on the intersection of politics and computer-mediated communication. She also serves as a commissioned public affairs officer in the U.S. Navy Reserve and is currently attached to U.S. Naval Forces Central Command/U.S. Fifth Fleet in the Middle East. She is an alumna of the Oxford Internet Institute's Summer Doctoral Programme and holds degrees from the University of Florida, Old Dominion University, and Tidewater Community College. She received an accreditation as public relations professional by the Public Relations Society of America in 2006.

LUCILA VARGAS is an associate professor in the School of Journalism and Mass Communication at the University of North Carolina at Chapel Hill. She holds a licenciatura degree from the Universidad Autónoma de Chihuahua, and an M.A. and Ph.D. from the University of Texas at Austin. Her teaching and research are in international communication, Latino media studies, and communication for social change. She wrote *Social Uses and Radio Practices: The Use of Radio by Ethnic Minorities in Mexico* (Westview, 1995), edited the anthology *Women Faculty of Color in the White Classroom: Narratives on the Pedagogical Implications of Teacher Diversity* (Lang, 2002), and has published in numerous academic journals.

JOHN YEMMA is a deputy managing editor of the *Boston Globe.* Since the beginning of 2006, he has been in charge of the *Globe's* transition to new media. For a year and a half leading up to the 2004 presidential election, Yemma was the *Globe's* national political editor, coordinating the work of editors and reporters in Boston and Washington and on the campaign trail. From September 2001 to 2003, Yemma was Sunday editor of the *Globe;* he played a role in Page One selection and special coverage of 9/11, the invasions of Afghanistan and Iraq, the Catholic Church abuse scandal, and other major stories during that period. In 1999, Yemma and his colleague Dan Golden won the Headliner's Award for beat coverage for their examination of Harvard University. Yemma was editor of the 1999 Polk Award–winning series on abuse of mental patients by medical researchers. He has received several *Globe* and *New York Times* awards for his contributions. From 1991 to 1996, Yemma was the *Globe's* foreign editor. He was a foreign correspondent in the Middle East in the early 1980s and covered wars in Lebanon and the Persian Gulf. He has also worked in various posts for the *Dallas Morning News,* UPI, and the *Christian Science Monitor.* In 1994 he was a Reuter Fellow at Oxford.

# INDEX